Principles of
Cell Proliferation

John K. Heath

School of Biosciences
University of Birmingham
Edgbaston
Birmingham

Blackwell
Science

© 2001
Blackwell Science Ltd
Editorial Offices:
Osney Mead, Oxford OX2 0EL
25 John Street, London WC1N 2BS
23 Ainslie Place, Edinburgh EH3 6AJ
350 Main Street, Malden
  MA 02148-5018, USA
54 University Street, Carlton
  Victoria 3053, Australia
10, rue Casimir Delavigne
  75006 Paris, France

Other Editorial Offices:
Blackwell Wissenschafts-Verlag
GmbH
Kurfürstendamm 57
10707 Berlin, Germany

Iowa State University Press
A Blackwell Science Company
2121 S. State Avenue
Ames, Iowa 50014-8300, USA

Blackwell Science KK
MG Kodenmacho Building
7–10 Kodenmacho Nihombashi
Chuo-ku, Tokyo 104, Japan

First published 2001

Set by Best-set Typesetter Ltd.,
Hong Kong
Printed and bound in Great Britain
by MPG Books Ltd, Bodmin,
Cornwall

The Blackwell Science logo is a
trade mark of Blackwell Science Ltd,
registered at the United Kingdom
Trade Marks Registry

A catalogue record for this title
is available from the British Library

ISBN 0-632-04886-7

Library of Congress
Cataloging-in-publication Data

Heath, J.K.
  Principles of cell proliferation /
J.K. Heath.
        p.   cm.
  Includes index.
  ISBN 0-632-04886-7
  1. Cell proliferation.
  2. Carcinogenesis.   I. Title.
  QH605 .H34   2000
  571.8′4—dc21

DISTRIBUTORS
Marston Book Services Ltd
PO Box 269
Abingdon, Oxon OX14 4YN
(*Orders*: Tel: 01235 465500
            Fax: 01235 465555)

USA
  Blackwell Science, Inc.
  Commerce Place
  350 Main Street
  Malden, MA 02148-5018
  (*Orders*: Tel: 800 759 6102
              781 388 8250
              Fax: 781 388 8255)

Canada
  Login Brothers Book Company
  324 Saulteaux Crescent
  Winnipeg, Manitoba R3J 3T2
  (*Orders*: Tel: 204 837 2987)

Australia
  Blackwell Science Pty Ltd
  54 University Street
  Carlton, Victoria 3053
  (*Orders*: Tel: 3 9347 0300
              Fax: 3 9347 5001)

For further information on
Blackwell Science, visit our website:
www.blackwell-science.com

Coventry University

Contents

Preface

This book aims to tell the story of how cells multiply. It is written for students, and others with some background in molecular biology who are coming to this problem for the first time. It tries to provide a scientific framework with which to understand new discoveries rather than a definitive and detailed report. This account is, however, no more than a snapshot taken from the current perspective. One of the most beguiling features of science is that you never truly understand anything. Be warned Reader!—there are numerous unanswered questions lurking in these pages and many confident statements that will soon be rendered irrelevant by new developments.

There are topics which have been excluded from this text and others which may well be dwelt upon too long. However, the personal view is intrinsic to the process and the way the book is written simply reflects my own attempts to comprehend the subject. I have, however, benefited enormously from the advice of others in this task. These include, not least, several generations of students who provided their views on various portions of the book at different stages of gestation. Anonymous reviewers read versions of the manuscript with great care and helped clarify my thoughts in many sections. Bob Michell gave it a stylistic workover. Chris Marshall, Andrew Wyllie and Michael Hengartner kindly provided images. Neil Hotchin provided the cover image. Anne Stanford and Delia Sandford provided tactful and patient editorial support and Helen Mardon sufficient helpful pressure to see the task completed. Most thanks must go, however, to the many colleagues in the field whose writings, thoughts and scientific findings have helped shaped my perspective over the years.

I have deliberately avoided ascribing credit to the scientists who made the discoveries. There are too many to name and they know who they are. The one exception is Robert Holley whose writings first fired my interest in this topic and to whom the book is gratefully dedicated.

John K. Heath
Birmingham

Abbreviations

ATP	adenosine triphosphate
BMP	bone morphogenetic protein
CDK	cyclin-dependant kinase
CHD	cytokine homology domain
DAG	diacylglycerol
DNA	deoxyribonucleic acid
ECM	extracellular matrix
EGF	epidermal growth factor
ERK	map kinase (MK)
FGF	fibroblast growth factor
FNIII	fibronectin type III
GAP	GTPase activating protein
G-CSF	granulocyte colony-stimulating factor
GEF	guanine nucleotide exchange factor
GM-CSF	granulocyte–macrophage colony-stimulating factor
Grb	growth factor receptor bound
GTP	guanosine triphosphate
HBEGF	heparin-binding epidermal growth factor
IGF	insulin-like growth factor
IL	interleukin
MAPK	mitogen-activated protein kinase
MDM	mouse double minute
MEK	MAP kinase kinase (MKK)
MIS	Mullerian inhibitory substance
MKKK	MAP kinase kinase kinase
MMTV	mouse mammary tumour virus
MPF	maturation promoting factor
NGF	nerve growth factor
NMR	nuclear magnetic resonance
PCNA	proliferating cell nuclear antigen
PDGF	platelet-derived growth factor
PDGFR	platelet-derived growth factor receptor
PH	pleckstrin homology
PI3K	phosphatidylinositol-3′-OH kinase
PKC	protein kinase C
PLC	phospholipase C
PTB	phosphotyrosine binding
RNA	ribonucleic acid

RTK	receptor tyrosine kinase
SRE	serum response element
SRF	serum response factor
STAT	signal transducer and activator of transcription
TCF	ternary complex factor
TGFA	transforming growth factor alpha
TGFB	transforming growth factor beta
TIMP	tissue inhibitor of metalloproteinases
TNF	tumour necrosis factor
TRE	TPA regulatory element
VEGF	vascular endothelial cell growth factor

Chapter 1: Biology of the Cell Cycle

Introduction

The defining feature of living organisms is the ability to multiply by replication of genetic material. This process is subject to strict controls. In the case of single-celled organisms such as bacteria the multiplication of the organism is closely linked to nutrient availability. For multi-celled organisms (such as ourselves) the proliferation of individual cells must be integrated with the overall needs of the organism and therefore subject to some form of coordination. This is achieved by subjecting the behaviour of individual cells to control by signals emanating from other cells.

The overall purpose of this book is to explain the molecular mechanisms that lie behind the decision of an individual cell to multiply. In this chapter we examine the basic biology of cell proliferation in order to define the primary features of cell multiplication for which a molecular mechanism should be sought.

The cell cycle

The multiplication of a single cell involves two essential processes: (i) the replication of deoxyribonucleic acid (DNA); and (ii) packaging the replicated DNA into two daughter cells by cell division. These two processes, in most circumstances, must be coordinated in time such that DNA replication is both initiated and completed before cell division occurs. There are some exceptions to this rule. The most important, which occurs in all sexually reproducing multi-celled organisms, is the production of haploid gametes and this occurs by partitioning of genetic material into daughter cells without prior replication. In some specialised cell types DNA replication can occur without cell division, leading to the production of cells with many times the normal content of DNA. These two exceptional circumstances indicate that DNA replication and cell division are independent processes which, under normal circumstances, are linked together. This orderly procession of DNA replication and cell division is called the cell cycle.

Phases of the cell cycle

Aside from defining the basic processes of the cell cycle, DNA replication and cell division are landmark events that can be readily observed and measured. DNA replication involves the incorporation of nucleotide precursors into DNA, which can be detected by exposing cells to radioactive or chemically labelled

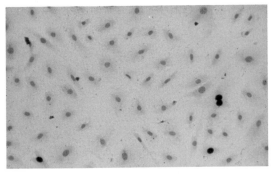

(a)

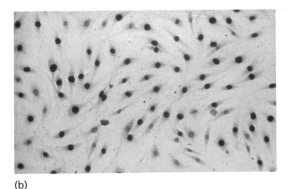

(b)

Fig. 1.1 Detection of cells in S phase by incorporation of radioactive nucleotide precursors. A population of cells in (a) G1 phase of the cycle and (b) S phase were exposed to radioactively labelled nucleotide precursors for a period of time. The cells were then fixed, and nucleotides that had not been incorporated into DNA were washed out. The cells were then exposed to film. Cells that were undergoing DNA synthesis in the course of the experiment have black nuclei.

nucleotide precursors (Fig. 1.1). Once the process of DNA replication is complete a cell has twice the DNA content it had before. A cell that has undergone DNA replication can therefore be detected by examination of its DNA content. The process of cell division in eukaryotic cells (mitosis) involves the condensation and orderly partitioning of pairs of chromosomes followed by cell fission into two daughter cells. Cells undergoing mitosis can be readily detected in the light microscope by virtue of their characteristic condensed genetic material and the possession of the mitotic spindle apparatus required for chromosome partitioning (Fig. 1.2). Once a cell has undergone cell division two daughter cells exist, where one mother cell existed before. The occurrence of cell division can therefore be inferred by measurement of the number of cells present in a particular population over time.

Kinetics of the cell cycle

Armed with the experimental means to measure and detect DNA replication and cell division, it becomes possible to study the kinetics of cell proliferation in a population of cells. It will be immediately apparent that it is not easy to measure the timing of DNA synthesis and cell division simultaneously in all cells of a population. In most cases it is necessary to bring together a variety of methods

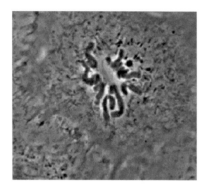

Fig. 1.2 A cell undergoing mitosis. This frame is taken from a video film of cells multiplying in culture. (Courtesy of Ted Salmon, University of North Carolina.) Mitosis involves highly characteristic morphological changes including: breakdown of the nuclear envelope, chromosome condensation, alignment of chromosomes on the spindle; and, finally, separation of homologous chromosomes and cell fission.

and build up a composite picture of the timing and duration of cell cycle phases in a growing population. There are three basic combinations of experimental techniques that can be used to define the kinetics of cell multiplication.

The first involves sampling a population of cells to determine the proportion which are undergoing either S phase or mitosis at the time of sampling, or by measuring DNA content, to determine that proportion of the population which have undergone DNA synthesis. The second approach involves blocking cells once they reach a landmark event and measuring the proportion of cells stuck at a particular landmark over time. The completion of mitosis can be blocked by the application of drugs, such as colcemid, which dissolve the mitotic spindle. The completion of DNA synthesis can similarly be blocked by drugs, such as hydroxyurea, which block the process of DNA synthesis itself. A variation on this approach is to block cells at one landmark for a defined period of time and then release the block and measure the time taken to reach the next landmark.

The final approach, and perhaps intuitively most satisfying, is to observe a population of cells continuously over a period of time using, for example, time-lapse video recording and to simply record when cell division occurs. It will be apparent that each of these approaches has its advantages and disadvantages but by pooling data from each approach it is possible to make some general statements about the duration of cell cycle phases and the kinetics of cell proliferation within a population.

The first important finding is that, in general, DNA synthesis and cell division are separated in time. Upon the 'birth' of a cell a period of time elapses before DNA synthesis is initiated. This 'gap' in the cell cycle is called the G1 phase. It also emerges that another period of time elapses after a cell has completed DNA replication but before it enters mitosis. This second gap is called the G2 phase of the cell cycle. The presence of these gaps, in which neither DNA synthesis nor mitosis is taking place, suggests the existence of two extra phases of the cell cycle in addition to DNA synthesis and mitosis. Combining these two gap phases with the landmark phases of DNA synthesis (or S phase) and mitosis (or M phase) leads to the definition of the cell cycle as the sequential activation and completion of four phases: G1, S, G2 and M. The multiplication of a popula-

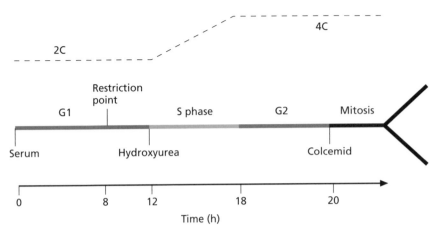

Fig. 1.3 The key events in the cell cycle. S phase and mitosis are the landmark events of the cell cycle which can be blocked by drugs such as hydroxyurea and colcemid. The DNA content doubles during DNA synthesis. The restriction point is placed before S phase during the G1 phase of the cycle.

tion of cells from a single founder can now be viewed as the quasi-repetitive ordering of four events in time (Fig. 1.3).

Having established the existence of four defined and ordered phases it becomes possible to analyse their duration in different populations of cells. From this type of analysis a second significant principle emerges: the duration of the cell cycle is not fixed but, on the contrary, is highly variable in its overall duration. Within this general pattern of variability significant differences between the four phases can be discerned. M phase is typically rapid and is often initiated and completed in less than an hour. The duration of M phase also shows little variation in timing between populations of cells multiplying at different rates. S phase is also fairly constant in duration in different populations of cells, typically being completed within four to five hours. There are some notable exceptions to this picture, especially in the early cell divisions following fertilisation of some species, when the S phase is condensed into a period of less than an hour. G2 is also generally constant in duration, although, again, some important exceptions are seen in early embryos where G2, like S phase, is highly condensed in duration or, conversely, in unfertilised eggs where G2 is indefinitely extended in duration until it is abruptly terminated upon fertilisation. Despite these exceptions the general cause of variability in the duration of the cell cycle, even in apparently identical cells, is the duration of the G1 phase. This can be graphically illustrated by examining the duration of cell cycle phases in populations of cells expanding at different rates (Fig. 1.4). This type of analysis reveals that the major determinant of cell cycle duration is the length of the G1 phase. The variability of the G1 phase is not due to simple stochastic processes. For any population of cells growing at different rates there exists a minimum duration of G1. As

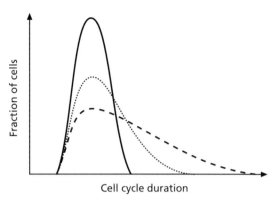

Fig. 1.4 Cell cycle duration in populations of cells multiplying at different rates (fast, solid line; medium, dotted line; and slow, dashed line). As overall population doubling rates decrease, the duration of cell cycles within the population becomes more variable.

population doubling times increase it is the variability of G1 that increases rather than its absolute duration (Fig. 1.4). This finding clearly indicates that G1 phase involves the activation of some process that is inherently variable in duration and which thereby determines the rate at which a population of cells increases.

Senescence

Kinetic analysis of the cell cycle has revealed the orderly progression of four defined periods of varying span and pinpointed G1 as the determining phase of cell cycle timing. A further significant feature of the cell cycle in multicellular organisms can be revealed by examining cell cycle duration as a function of the historical behaviour of a population of cells. If a population of cells is followed over a period of time it can be seen that not only does the population doubling time become progressively more variable with time but, in particular, a significant faction of cells fails to complete the cell cycle and give rise to progeny cells at all. These cells are termed 'senescent'. Cellular senescence is characterised as a process by which cells enter the G1 phase of the cycle but fail to complete G1 and never enter S phase again. As time passes the fraction of senescent cells in a population increases until there comes a point at which a 'plateau phase' occurs and further increase in the population is halted entirely (Fig. 1.5). The total number of progeny cells that can be generated from a founder cell is termed the proliferative potential. It follows that many cell types in multi-celled organisms have a finite proliferative potential. It will immediately become clear that senescence is a highly significant biological phenomenon, since many cell types in the body, such as neurons, muscle cells and red blood cells, are effectively senescent in that they never undergo cell division at all, although they were derived from proliferating parental cell types.

The phenomenon of cellular senescence exhibits some intriguing features. Whilst the proliferative potential of a particular cell type is fairly constant

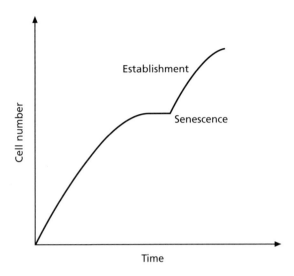

Fig. 1.5 Senescence and establishment. Cells are allowed to multiply in culture for a prolonged period of time and the size of the population is monitored. The rate of population increase begins to slow after a time and eventually no further increase occurs. At this point the cells are senescent. After another period of time the population size starts to increase again. This is the result of a fraction of cells in the population undergoing establishment and re-initiating cell multiplication.

between individuals, it is highly variable between different cell types. This means that it is possible to define a population of cells by its proliferative potential. This is most dramatically illustrated by examination of certain cell types in the early embryo which may undergo senescence after only a few cell divisions (Fig. 1.6). In other cases the proliferative potential of cells may exceed that which normally occurs during the lifetime of an organism. In other words proliferative potential (or the propensity of cells to enter senescence) is a 'programmed' feature of cell type rather than a simple random loss of the capacity to divide.

Close inspection of a population of cells approaching senescence reveals that the proliferative potential of individual cells varies widely within an otherwise identical population. Individual cells within a population may give rise to widely differing numbers of progeny before eventually succumbing to senescence. There are some analogies with the duration of the G1 phase in this phenomenon, in that the overall rate of population expansion conceals significant heterogeneity in the behaviour of individual cells within the population. In both cases the key regulatory event is manifested in the characteristics of the G1 phase of the cell cycle.

Quiescence

The key feature of senescence is that it is an irreversible event. Cells may, however, cease to proliferate but retain the capacity to re-initiate progress through the cell cycle at a later date. This phenomenon is called 'quiescence'. Quiescence can be induced in many types of cells by manipulation of their environment. The means by which this is achieved, and its significance for dissecting cell cycle control mechanisms, will be discussed later. Quiescence represents an impor-

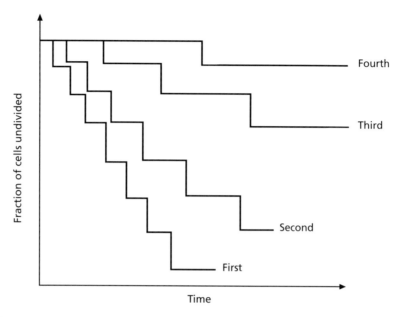

Fig. 1.6 Hypothetical example of cell senescence. Individual cells from the culture described in Fig. 1.5 are monitored continuously and cell cycle duration of the first, second, third and fourth rounds of division is measured. As cells approach senescence, the cell cycle duration increases in variability and the fraction of cells that fail to give rise to further progeny increases.

tant level of biological control for the whole organism since it indicates that cell multiplication may be held in check under normal circumstances, but rapidly induced when required, as for example in the expansion of immune effector cells during infection or in tissue regeneration following injury.

Quiescence exhibits many parallels with senescence, not least the fact that quiescent cells remain in the G1 phase of the cell cycle. Moreover, when cells exit quiescence they remain in the G1 phase of the cell cycle for some time before the onset of S phase (Fig. 1.7). The kinetics of entry into S phase following the exit from quiescence are very similar to the variability of the G1 phase under normal conditions of continuous cell proliferation. This suggests that at least part of the variability of the G1 phase arises from events that occur late in the G1 phase and that quiescence involves a regulatory event that precedes the event determining the duration of G1. This has led to the idea that quiescent cells have entered an 'extra' phase in the cell cycle, preceding G1, termed G0. This idea proposes that at the beginning of the life of each cell there is an option either to initiate the processes of G1 or to enter G0. The decision to enter G0 is reversible, since quiescence can be relieved by alteration of the cell's environment. In this light the phenomenon of senescence may reflect an inability to exit G0.

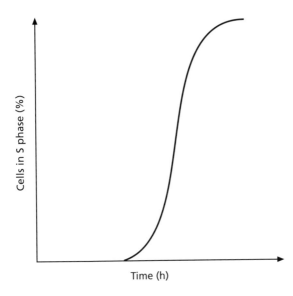

Fig. 1.7 Induction of DNA synthesis in quiescent cells involves a period in G1 before DNA synthesis begins. Cells are rendered quiescent and then stimulated to re-initiate cell cycle progression by the addition of serum. The fraction of cells in S phase is monitored over time. There is a minimum time before the first cells entering S phase are detected and variability in the rate at which cells in the population enter S phase.

Establishment

Taking all the kinetic evidence together the clear idea emerges that the G1 phase of the cell cycle is not an 'empty' gap that simply functions to separate different processes in time but a phase in which the future behaviour of a cell, in terms of the completion of the cell cycle, is determined. The existence of senescence and quiescence clearly indicate that molecular mechanisms exist that govern cell proliferation in the G1 phase of the cell cycle by determining whether or not a cell will divide in future. There are, however, a set of circumstances in which it is possible to experimentally separate the two phenomena. If certain types of cells are grown for prolonged periods of time, their proliferative potential is realised and further population expansion is halted. Over time, however, clonally-derived subpopulations of cells can arise which re-initiate cell multiplication and the population starts to expand further (Fig. 1.5). These subpopulations of cells have lost the capacity to enter senescence and are termed 'established'. Established cells are, in effect, immortal since they have the capacity to proliferate indefinitely and therefore have an infinite proliferative potential. Established cells have not, however, lost the ability to undergo quiescence. Therefore, whatever events lie at the heart of establishment, they seem to influence the process of senescence but not quiescence.

Establishment has a number of curious features. The clonal origins of established cells indicate that the process involves some rare event that may not necessarily be executed in the lifetime of an organism. Establishment does not, therefore, represent some normal physiological control process but rather some anomalous manifestation of cell behaviour.

The essential abnormality of establishment is supported by the fact that it is invariably accompanied by genetic alterations and, in particular, the acquisition of genome instability. Established cell types often exhibit abnormal numbers of chromosomes and other types of gross genome rearrangements. This indicates that the processes underlying establishment may involve some form of genetic alteration to the cell.

The propensity to give rise to established cell types is very much a feature of the species of origin. Different cell types also differ widely in their ability to give rise to established cell types within the same species. Thus, established cell 'lines' (so-called after their clonal origins) can be established relatively readily from connective tissue fibroblast cells of rodents but almost never from humans or chickens. On the other hand, it is very difficult to generate established cell lines from many other types of cells in rodents. The propensity to establishment therefore also exhibits some form of species- and cell-type-specific programming suggesting that, if establishment has a genetic origin, multiple target genes may be involved.

Whilst establishment may be an essentially abnormal phenomenon, it has enormous practical value for the investigator of cell cycle control mechanisms. Established cell lines represent a population of cells whose behaviour (subject to the absence of further genetic changes) is uniform over time and, by virtue of their retained ability to enter and exit quiescence, can be experimentally manipulated in their cell cycle behaviour. Established cell lines such as mouse 3T3 cells have therefore become the 'workhorse' experimental material for many of the investigations of cell cycle control to be described below.

Stem cells

An important concept to emerge from a consideration of cellular proliferative senescence and quiescence is the idea that the proliferative potential of a particular cell is developmentally programmed. In other words, the ability of a cell to progress through the cell cycle and the potential number of its future progeny is dependent upon the exact nature of the cell in question. It follows that cells with a limited proliferative potential must be derived from cells with a greater proliferative potential. This could occur in the course of the normal life history of a particular population of cells or by differentiation from a precursor cell type with enhanced proliferative capacity. Perhaps the most extreme example of the latter case would be the fertilised egg which, by definition, under normal circumstances has the greatest proliferative potential of all the cells in the body. This idea has particular attraction when considering tissue systems that exhibit substantial capacity for continuous or intermittent renewal. Examples of such systems include the gut, skin and haemopoietic system which not only are undergoing continuous turnover and self-renewal in the course of normal life but also exhibit remarkable regenerative capacities when subjected to massive cell loss. The ability of such systems to sustain immense production rates of mature cells has led to the concept of the 'stem cell'.

The stem cell concept is somewhat slippery and has been interpreted to mean different things by different investigators. Strictly speaking, it refers to a situation in which a single cell may divide to give two progeny, one of which retains the biological properties of the parent and the second of which differs in some physiologically significant respect. In the context of tissue systems undergoing continuous self-renewal the concept has been used to suggest that at the root of such systems is a cell type with a very high, perhaps infinite, proliferative potential. Upon division this cell type may retain its original state or give rise to progeny cell types with reduced and, importantly, finite proliferative potential. Under normal circumstances it is thought that such stem cells are either quiescent or slowly proliferating and the majority of the tissue population is sustained by proliferation of the more mature progeny cell types. If the mature progeny cells are eradicated the stem cell population expands in number and this provides the means to rapidly repopulate a tissue compartment.

The evidence for the existence of such cell types is, for the most part, indirect and relies on two types of information. The first is that tissue systems such as the gut are able to regenerate following massive cell loss by agents, such as radiation or cytotoxic drugs, which eliminate the majority of dividing cells in the population. This implies that the entire population of dividing mature cell types in such tissues is derived from a precursor which, under most circumstances, is progressing through the cell cycle too slowly to be affected by the cytotoxic insult.

The second line of evidence involves the serial transplantation of cells into hosts in which a self-renewing tissue compartment has been eliminated. This type of experiment has, in particular, been performed in the haemopoietic system, where it has been possible to demonstrate that repopulation occurs by expansion of a minute subset of cells present in the entire system. This tiny subpopulation has the capacity to withstand serial transplantation and therefore clearly has a proliferative potential that is never normally required within the lifetime of an individual organism. The main problem with this concept has historically been the difficulty of isolating individual cells with stem cell properties. It is nevertheless clear that developmental regulation of the proliferative potential by programming the rate of cellular senescence may play a very important role in normal and pathological tissue physiology.

The growth of cells *in vitro*

It is clearly desirable to analyse the molecular controls on cell proliferation under tightly controlled experimental conditions. Cells in tissues, whilst representing the physiological ideal, are difficult to manipulate experimentally and, for reasons described above, are often highly heterogeneous in terms of their past or future behaviour. For this reason much attention has been devoted to analysing the multiplication of cells *ex vivo* in culture (*in vitro*). This has proved to be a key informative experimental approach to the molecular dissection of cell cycle control mechanisms, although, as will be documented frequently below, the 'simplicity' of *in vitro* models of cell behaviour can be misleading when attempts are made to extrapolate findings back to real cells in real organisms.

The requirements for proliferation *in vitro*

The first step in analysing cell proliferation *in vitro* is to establish conditions under which cell populations can be made to grow in the culture dish. Experimental approaches to this issue have been made since the early 1900s and for many years largely involved a considerable amount of guesswork and 'trial and error' approaches. The problem was also complicated in these early studies by the requirement to work with heterogenous cells derived directly from normal tissue. The discovery and development of established cell lines, described above, removed many of these difficulties since they were a population of cells of uniform origin which, unlike their normal counterparts, were not destined to cease proliferation in the course of an experiment and could therefore be analysed over long periods of time. The most popular subject of these investigations was established cell lines (such as 3T3) of fibroblastic origin which grow as a monolayer attached to a plastic (or glass) surface (Fig. 1.8).

Essentially three categories of additives are required in the growth medium of cells to obtain continuous proliferation *in vitro*. The first are compounds which could be generically categorised as nutrients. These include energy sources such as glucose or pyruvate as well as amino acids, vitamins and micronutrients, such as sources of iron, cobalt, manganese and selenium. An

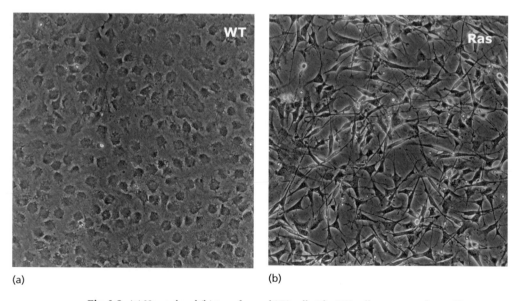

(a)   (b)

Fig. 1.8 (a) Normal and (b) transformed 3T3 cells. The 3T3 cells were transformed by expression of a mutant *Ras* gene (see Chapter 7). Note the dramatic change in cell morphology and breakdown of a tight cell monolayer. (Pictures courtesy of Hugh Paterson and Chris Marshall, Institute for Cancer Research.)

optimal balance of nutrients is essential for cell viability but does not, by itself, have any effect on cell cycle kinetics *in vitro*.

The second category of additives are molecules required for the cells to adhere to the solid substratum, such as fibronectin, collagen and vitronectin. (These elements are obviously not required for cells that proliferate in suspension, such as those of lymphoid origin.) For adherent cell types, such as established fibroblast cell lines, attachment to a substrate is a necessary pre-requisite for cell proliferation but is not essential since cells may remain attached to the substrate yet fail to proliferate. In order to multiply in culture (as opposed to merely survive), an additional category of additives is required and these have traditionally been added in the form of animal serum.

The discovery of serum as a source of additives required for cell multiplication finds its origins in the early attempts to make primary cells grow *in vitro*. Investigators found that certain tissues would undergo limited proliferation *in vitro* under conditions designed to emulate the conditions that might occur at the site of wound: namely, in the presence of clotted plasma or serum. The significance of the discovery of serum as a source of 'growth factors' was that it provided a method for experimentally manipulating cell proliferation in culture and thereby opened the way to the biochemical analysis of cell proliferation.

The classic experiments on the role of serum in the control of cell proliferation were performed by Robert Holley and colleagues. They studied the growth of the established fibroblast cell line 3T3 in culture. It was first noted that the multiplication of 3T3 cells was determined by the concentration of serum present in the culture. If equal numbers of cells were plated in different concentrations of serum, they proliferated for a period of time and then entered quiescence in the G1 phase of the cell cycle. The number of population doublings that took place was directly related to the original concentration of serum present (Fig. 1.9). Cells that entered quiescence in low concentrations of serum could, however, be induced to re-enter the cell cycle and undergo further proliferation on the addition of fresh serum to the culture. The number of extra cell cycles that occurred was directly related to the concentration of serum added and closely corresponded in number to that of cells that had been originally plated in higher concentrations. The conclusion from these, and related experiments, is clear. The proliferation of cells is dependent upon factors that are present in serum and which are depleted with continued cell proliferation. Upon exhaustion of these serum-derived growth factors, the cells enter quiescence in the G1 phase of the cell cycle until stimulated to undergo further rounds of cell division by the addition of fresh serum. In other words, exit from quiescence is induced by exogenous factors that can be provided in the form of serum.

The restriction point

The recognition that serum could be empirically employed to control entry into and exit from the cell cycle opened the way to further investigation of the

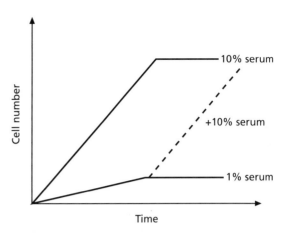

Fig. 1.9 The relationship between serum and cell proliferation of 3T3 cells. Populations of 3T3 cells are exposed to different amounts of serum (1% and 10%) and cell numbers monitored over time. The cells in 1% serum undergo fewer rounds of division than those in 10% serum. Cells in 1% serum can be induced to undergo further rounds of division by exposure to fresh serum. The final population size reflects the amount of serum to which the cells have been exposed. (Redrawn from Holley, R.W. & Kiernan, J.A. *Proceedings of the National Academy of Sciences* **71**, 1286.)

role of these exogenous serum-derived factors in the control of cell proliferation. In particular, it was possible to ask whether these serum-derived factors were required to be present for the whole or part of the cell cycle. Thus, quiescent cells could be stimulated with serum that was subsequently withdrawn at varying periods of time thereafter. These experiments showed that serum was required only for part of the cell cycle corresponding to a major fraction of the G1 phase. Once the serum-dependent point in G1 had been passed, the cells were able to execute the remainder of the cell cycle in the absence of exogenous factors. This point in G1 where completion of the cycle became independent of exogenous factors was termed the 'restriction point'. The timing of the restriction point in G1 was intriguingly similar to that point at which progress through the cell cycle becomes sensitive to inhibition by drugs that block protein synthesis, suggesting that the empirically defined restriction point corresponds to some form of biochemical event that is dependent upon protein synthesis for its execution.

The general conclusions from these experiments are clear. The exit from quiescence and progress through the early part of G1 is induced by exogenous factors that can be conveniently supplied in the form of animal serum. These factors appear to induce progress through G1 until the restriction point is reached, at which time the cell is committed to complete the rest of the cell cycle. The G1 phase of the cell cycle therefore seems to involve at least two critical regulatory events: the decision to exit quiescence and the decision to complete G1, S, G2 and M (Fig. 1.3).

Cell size

It is very clear that the process of progress through the cell cycle has to involve a net increase in cell mass. If not, daughter cells would be smaller than their parents, leading to ultimate catastrophe. Indeed, ongoing protein synthesis is required for cells to complete progress through the cycle and a requirement for exogenous biosynthetic precursors is evident throughout the cell cycle. In some unicellular organisms (described further in Chapter 6) there is a strict relationship between the size of a cell and its position in the cell cycle. In multi-celled organisms, however, the relationship between cell size and cell cycle appears more complicated. Part of the problem is that it is difficult to rigorously define what is meant by cell size: is it the area occupied by a cell? its volume? or net mass? Using any of these criteria, it is evident that different cell types in the body have widely differing 'masses'. In addition, cells within an otherwise uniform population appear to have differing sizes and shapes (look closely at Fig. 1.8). Experiments aimed at measuring the relationship between the volume of cells and their position in the cycle fail to give any strict correlation between the two parameters. This indicates that observable measures of cell size may indirectly reflect some underlying process such as the rate of protein synthesis. In this light it is clear that all cells, whatever their mass, will require an approximate doubling of protein content in the course of one cell cycle but the net amount of protein required may vary from cell to cell. This indicates that part of the mechanism of progress through the cell cycle must involve a coupling of cell cycle events to general protein synthesis. The cell cycle therefore involves coupling progression to the translation of specific proteins (as manifest at the restriction point) as well as to general protein biosynthesis.

Cell transformation and tumours

So far we have been concerned with the behaviour of 'normal' cells derived from normal tissues. As is the case for much of biology, however, key insights have been obtained from the investigation of abnormal phenomena. In particular, it had been recognised for many years that many of the regulatory cell cycle phenomena described above did not pertain in the case of cells derived from tumours. The defining feature of a tumour is that cells proliferate under conditions where their normal counterparts do not. It is the progressive expansion of a population of tumour cells that gives rise to the central pathological consequences of malignant disease. An important practical point about cells derived from tumours was that, in many cases, they proved much easier to propagate *in vitro* than their normal counterparts and their cell cycle behaviour could therefore be investigated in some detail. The main lesson to emerge from these studies is that tumour-derived cells break many of the rules defined from the study of normal cells.

It is important to appreciate that the process of changing from a normal cell to a malignant tumour cell involves multiple changes, usually of genetic origin; tumour cells exhibit many, and varied, differences to their normal counterparts. This issue will be developed more fully in Chapters 7 and 8. In terms of cell cycle control, however, a number of essential differences can be defined.

The first and most obvious difference is that tumour cells behave like established cell lines in that at least part of the population is unable to undergo senescence. It is this feature, above all, which results in the progressive growth of the tumour *in vivo*. Established cell lines do not, however, give rise to tumours. It follows that additional differences must exist between tumour cells and established cell lines such as 3T3. It became possible to study these differences in more detail when cell lines were derived from 3T3 cells, by either treatment with known carcinogens or infection with tumorigenic viruses, whose behaviour *in vivo* exactly matched that of a naturally occurring tumour. Such cell lines were termed 'transformed' and the existence of transformed variants of established cell lines permitted, for the first time, a reasonable (although not necessarily exact) comparison between tumorigenic and non-tumorigenic cells of common origin.

A clear difference between normal and transformed 3T3 cells is apparent upon visual inspection (Fig. 1.8). Whereas normal fibroblasts exhibit a flat 'cobblestone' morphology, transformed 3T3 cells are characterised by a more elongated phenotype and typically grow on top of each other to form characteristic foci. The appearance of multi-layered foci forms an experimentally convenient method of identifying and selecting transformed cells from their normal counterparts. This relaxation of a strict requirement for attachment to a solid substrate is most clearly manifest in the phenomenon of 'anchorage-independent growth', whereby transformed 3T3 cells are able to grow in suspension in semisolid media such as agarose and form tight colonies. It is important to appreciate, however, that these morphological differences are not absolute but can be transiently emulated by normal 3T3 cells under appropriate experimental conditions.

The most significant manifestation of the transformed phenotype becomes apparent when transformed 3T3 cells are subjected to the experimental regimen devised by Holley and colleagues, described above. The proliferation of transformed cells exhibits a striking difference in the requirement for serum-derived factors for cell proliferation. Thus, transformed 3T3 cells are able to proliferate in low concentrations of serum where the growth of normal counterparts would be rapidly restricted (Fig. 1.10). Indeed, a defining feature of cell transformation is a significantly reduced and, in many cases, complete absence of a requirement for exogenous serum-derived factors to sustain progress through the cell cycle. It follows from this that the process of cell transformation involves some alteration to the biochemical mechanisms which underlie the two regulatory events that occur in the G1 phase of normal established cell lines, in addition to a loss of the ability to undergo proliferative senescence that is characteristic of cells in normal tissues. The reason why transformed cells give rise to progressively growing tumours *in vivo* appears therefore, in part, to result from genetic changes which either side-step or constitutively activate biochemical checkpoints that occur in their normal counterparts.

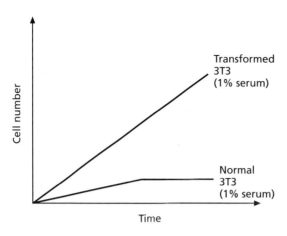

Fig. 1.10 Transformed cells do not require serum to proliferate. The experiment shown in Fig. 1.9 is repeated using normal and transformed cells exposed to 1% serum. The transformed cells reach higher population densities than their normal counterparts. (Redrawn from Holley, R.W. & Kiernan, J.A. *Proceedings of the National Academy of Sciences* **71**, 1286.)

Apoptosis and cell death

Thus far we have considered the multiplication of a population of cells to be entirely a function of progression through the cell cycle. This is a biologically simplistic view that ignores the impact of cell survival on population increase. For many years it had been assumed that cell death simply resulted from metabolic failure of some kind. In particular, the frequent observation of massive cell death occurring when normal cells were placed in culture was assumed to arise from an inappropriate mixture of environmental nutrients or the presence of noxious elements in the culture milieu. In this view, cell death could be considered a passive and avoidable phenomenon of little physiological consequence.

This view required radical reappraisal when the extent of cell death in normal tissues, and its significance for the control of population size, came to be appreciated. In particular, careful study of cell numbers in specific tissues such as the nervous system revealed an extensive loss of cells reproducibly occurring in the course of normal development. The essential 'normality' of cell death as a biological control on the size of cell populations was reinforced with the finding that specific loss-of-function mutations could inhibit cell death in a number of experimental systems. The implication of this finding was that cell death was an active biochemical process that required the participation of a specific set of gene products. This was further supported by the seemingly paradoxical finding that, in a number of *in vitro* systems, cell death could be inhibited by the pharmacological inhibition of gene expression or protein synthesis. The conclusion from these findings was that cell death was a specific regulatory event, an alternative to progression through the cell cycle, which was governed by specific biochemical mechanisms. The phenomenon of active programmed cell death has been termed apoptosis. The anatomical and molecular details of apoptosis will be described further in Chapter 9.

Conclusion

This chapter has described the essential features of the control of cell multiplication. Cell population size is controlled by the proliferation of cells, the onset of senescence and programmed cell death. The process of cell proliferation involves a series of events that occur in a reproducible sequence. The decision to complete the cell cycle involves a number of regulatory events that occur in the G1 phase, some of which are contingent upon the cell receiving exogenous signals. The process of malignant transformation involves many genetic changes, including some which act to bypass normal G1 control points. In the chapters that follow we shall start to analyse the molecular basis of these key concepts.

Chapter 2: Growth Factors

Introduction

The G1 phase of the cell cycle involves a biochemical process that commits the cell to progress the remaining phases of the cell cycle. This event is, under normal circumstances, dependent upon the action of external signals. In view of their role in controlling cell multiplication, these signalling molecules have been termed 'growth factors'. In this chapter we shall consider the biochemical identity and biological properties of growth factors.

General properties

Growth factors are a highly heterogeneous group of molecules whose shared feature is an ability, in specific circumstances, to regulate cell proliferation. Within this broad generalisation some common themes emerge. Growth factors are, for the most part, secreted proteins which are released from cells to act upon their target. A large number of molecules with growth factor activity have been defined which exhibit widely varying biochemical properties. Despite this variety it is possible to identify subgroups that can be conveniently linked together. These subgroups are of two types. Growth factor families are sets of molecules linked together on the basis of shared amino acid sequence features. The concept of growth factor families is not only useful as a simplifying abstraction but also has some biological rationale, since members of particular growth factor families often exhibit shared features in terms of their biological activity and biochemical mechanism of action. As an increasing number of growth factors have been characterised, and their three-dimensional structures determined, a second level of organisation has become apparent. This is the concept of 'super-families'. Growth factor superfamilies are sets of molecules which can be grouped together on the basis of shared features of tertiary structure. These shared structural features do not necessarily indicate shared biochemical actions (although this is often the case) but rather suggest that a set of molecules with highly disparate biological functions may have evolved from a much more restricted set of precursor types. The existence of a large repertoire of growth factors, derived from a limited set of common precursors, may therefore reflect both the complexity of cell types and complex physiological functions found in higher multicellular eukaryotes such as humans.

An important application of this idea is the notion that the diversity of molecules with growth factor activity found in nature reflects a means employed to achieve target cell specificity. Indeed, as discussed below, it is a fundamental property of all growth factors that they act only upon a characteristic and defined set of target cell types. There is, in other words, no 'universal growth factor' with identical actions on all cells but rather a repertoire of agents, each of which has a designated biological specificity.

Another important general feature of growth factor action is that, in most cases, the factors act locally within tissues rather than systemically. In this respect they can be distinguished from classical endocrine hormones such as insulin or prolactin. Such a local action makes biological sense when considering mechanisms by which the harmonious growth of a tissue, comprising multiple cell types in defined proportions and spatial relationships, may be achieved. The concept of local action has been further elaborated to define different modes of delivery of a growth factor to its target cell (Fig. 2.1). 'Endocrine' refers to the mechanisms employed by classical hormones, which are released from a specific gland or tissue into the general circulation to reach their target cell type. 'Paracrine' refers to the situation in which the growth factor is expressed immediately adjacent to the responsive target cell. 'Juxtacrine' is an extension of this to the situation in which the expressing cells and a target cell are in direct physical apposition. This requires that the growth factor is not released from the expressing cell but rather tethered by some mechanism. Finally, 'autocrine' describes the situation in which the expressing cell and the target cell are one

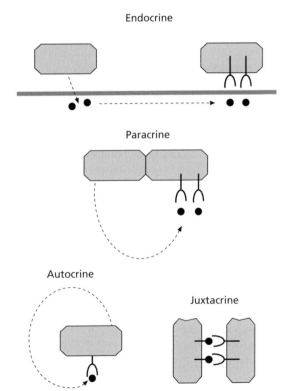

Fig. 2.1 The local action of growth factors. In endocrine systems a growth factor is released from a tissue and passes through the bloodstream to reach the target organ. In paracrine systems the growth factor is released from the cell and acts upon its close neighbours. In a juxtacrine system the growth factor is tethered to the cell and can only act upon another cell with which it has physical contact. In an autocrine system the growth factor acts upon the cell in which it is expressed.

and the same. It will be seen immediately that autocrine action of a growth factor would manifest as a situation in which there would be no apparent need for exogenous mediators of cell proliferation. Although these terms are useful in drawing attention to the local action of growth factors, it should be appreciated that, in real tissues, it is very likely that a mixture of these different modes of delivery occur concurrently.

Whilst it is a convenient experimental simplification to consider a single growth factor acting on a single target type, this can obscure an important feature of growth factor action *in vivo*. This is that the biological action of an individual growth factor can be dependent upon the prior or concurrent action of another. Growth factors are, in effect, assembled into regulatory circuits of varying kinds. The evidence for this proposition will be discussed further below but it will be seen that it provides another layer of specificity in the biological control of cell proliferation within tissues. Using this concept it can be imagined how the same growth factors can play distinct roles in different settings, depending on the context in which they act.

Finally, the emphasis in this chapter is on the action of growth factors in dictating cell multiplication and thus, indirectly, population size. It is now very clear that, in many cases, this is but one class of biological response which can be executed as a result of growth factor action. Growth factors are, in fact, pleiotropic in their actions. Depending upon the target cell type, a growth factor may elicit a variety of additional cellular responses, including the induction of cell migration and differentiation as well as apoptosis or survival. This aspect is a key feature when considering the overall biological functions of growth factors in controlling the mass and form of tissues *in vivo*.

Having considered some of the defining characteristics of growth factors in general terms we now turn to examine the biochemical characteristics of a selected set of growth factor families. It would be beyond the scope of this book to discuss every growth factor currently known. The family concept is, however, useful in defining the main outlines of growth factor biochemistry and laying the ground for understanding their biological activities.

The platelet-derived growth factor family

It is useful to recall the experimental origins of growth factor research described in Chapter 1. It was empirically found that animal serum (namely the non-cellular component of clotted blood) was a useful source of growth factor activity. Attention naturally turned to the biochemical identity of the active ingredients in animal serum. An important breakthrough occurred when the activity of serum was compared with plasma. It was discovered that, for a variety of cell types, serum was significantly more potent than plasma in supporting cell multiplication (Fig. 2.2). The main difference between serum and plasma is the presence of proteins released as a consequence of platelet breakdown during the clotting process. It was discovered that the biological

Fig. 2.2 The presence of
growth factors in platelets.
Smooth muscle cells were
grown in the presence of
plasma or serum. The cells
undergo few divisions in
plasma. Adding an extract of
platelets to plasma resembles
the growth-promoting effects
of serum. (Redrawn from
Vogel *et al. Proceedings of the
National Academy of Sciences* **75**,
2810–2814.)

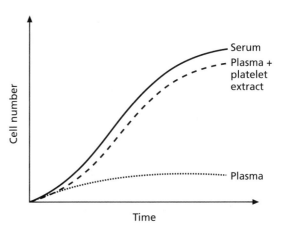

Cell number

Serum

Plasma +
platelet
extract

Plasma

Time

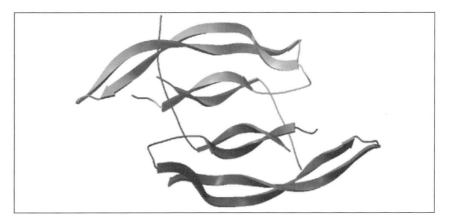

Fig. 2.3 The three-dimensional structure of platelet-derived growth factor (PDGF). The
protein is a disulphide-linked dimer of two anti-parallel chains.

activity of serum could be conferred upon plasma by addition of platelet extracts
(Fig. 2.2). This provided a relatively rich source of growth factor activity and
eventually led to the purification of the active agent. This was (naturally
enough) termed platelet-derived growth factor (PDGF). PDGF, as originally purified, was composed of two disulphide-linked subunits of molecular mass
12 500 kDa (Fig. 2.3).

Molecular cloning of PDGF genes, combined with purification of PDGF activity from alternative sources, led to the discovery that PDGF was actually derived from two genes, *PDGF-A* and *PDGF-B*, whose protein products could be
assembled via dimerisation into three distinct molecular forms: AA, BB and AB.
Various studies have indicated that these three molecular forms exhibit different biological activities when compared upon different target cell types. This is
most clearly illustrated when examining the effect of genetic inactivation of the
PDGF-A and *PDGF-B* genes in mice. These animals exhibit highly specific and

characteristic defects: *PDGF-B* deficient mice exhibit kidney dysfunction resulting from a loss of the glomerular mesangial cells, whereas *PDGF-A* deficient
mice exhibit lung dysfunction resulting from the loss of alveolar myofibroblasts.
These observations are surprising in that *PDGF-A* and *PDGF-B* exhibit widespread (and distinct) patterns of expression during embryonic and fetal life and
yet their presence is only critical under very specific circumstances. This is a
theme that will often reappear when considering the biological functions of different growth factors *in vivo*.

Close relatives of the PDGF family of growth factors are the vascular endothelial cell growth factors (VEGFs). Three members of the family are currently known: A, B and C. These are related in sequence to PDGF-A and -B, are also
expressed as disulphide-linked dimers, but exhibit an entirely distinct target cell
specificity. As their name implies, VEGFs are highly specific growth factors for
the endothelial cells of the blood vessel wall. Gene disruption experiments in
mice have indeed confirmed that VEGFs are absolutely required for the formation of different types of blood vessels during embryonic and fetal development.
An important feature of VEGF regulation is that expression is controlled by oxygen availability: under conditions of low oxygen tension, such as would arise in
the centre of a tumour or solid tissue, VEGF expression is induced and the ensu-
ing induction of endothelial cell multiplication causes new blood vessels to be
formed, thereby increasing oxygen availability to the tissue. This type of mechanism is an excellent example of how physiological regulation of tissue growth
in the adult animal is controlled by growth factors acting in response to environmental cues.

Both the PDGF and VEGF families are subjected to alternative splicing
mechanisms which yield two distinct forms of protein: 'long' and 'short'. The
long forms are characterised by the insertion at the C-terminus of a short
polypeptide sequence that is very rich in basic amino acids. This results in the
protein associating with the cell in which it was expressed as a result of binding
to both cell surface glycosaminoglycans and a specific 'anchoring' receptor,
neuropilin. This mechanism illustrates how the dissemination of growth factor
within a tissue can be subject to tight paracrine control. This concept will reappear frequently as we examine other growth factor families.

The epidermal growth factor family

Epidermal growth factor (EGF) was one of the first growth factors to be discovered. Injection of extracts of mouse submaxillary gland extracts into newborn
mice was found to result in accelerated maturation of various epithelia, leading
to premature eyelid opening and incisor eruption. EGF isolated from mouse
submaxillary glands is a 6-kDa polypeptide with mitogenic actions on a wide
variety of epithelial cells as well as 3T3 fibroblasts, in culture. Transforming
growth factor alpha (TGFA) was identified as an activity present in the culture
supernatants of virally transformed leukaemia cells that had the property of

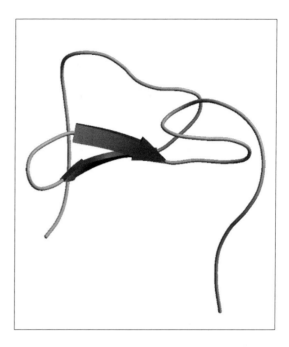

Fig. 2.4 The three-dimensional structure of epidermal growth factor (EGF). The protein has a core of two anti-parallel strands linked by loops. The EGF 'fold' is found in many proteins.

competing with EGF for binding to its cell surface receptor. The molecular cloning and characterisation of TGFA revealed a protein with homology to EGF which exhibited considerable overlap in biological functions. Both EGF and TGFA are expressed as transmembrane proteins from which the mature peptide is released by proteolysis. The solution structures of EGF and TGFA have been determined by nuclear magnetic resonance (NMR) techniques; this reveals a canonical fold, or module, that can be identified in many different proteins, most of which are devoid of mitogenic activity (Fig. 2.4). This indicates that the receptor recognition functions of the EGF family have evolved from a core structural framework that can be employed for other purposes.

In recent years the EGF family of growth factors has expanded in size (Fig. 2.5). Amphiregulin was purified from the culture supernatants of phorbol-ester-treated human breast adenocarcinoma cells as a growth inhibitor. Although amphiregulin shares considerable sequence homology with the mature forms of TGFA, it also had an extended N-terminus, rich in basic residues, which interacts strongly with heparan sulphate. This interaction of heparan sulphate is not only required for the biological activity of amphiregulin *in vitro* but also plays a role in the correct processing and folding of the nascent amphiregulin protein. Other EGF-like ligands have been identified; these include betacellulin, a growth factor secreted by experimentally induced pancreatic tumour cells, and heparin-binding epidermal growth factor (HBEGF). HBEGF is synthesised as a transmembrane protein and in this form it can act as the receptor for tetanus toxin. The biologically active ligand can there-

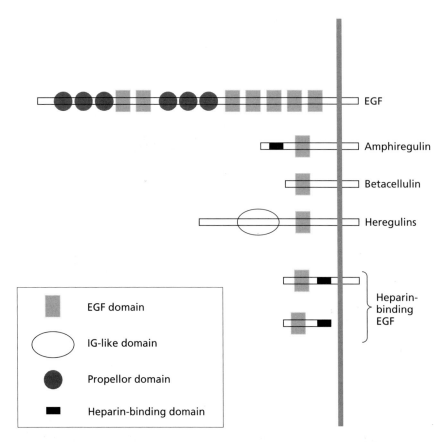

Fig. 2.5 The epidermal growth factor (EGF) family. The molecules are composed of differing numbers of structural elements but each contains a core EGF fold. All EGF precursors are transmembrane proteins.

fore either be presented to a responding cell in the form of a membrane-bound juxtacrine agent or be released from the cell membrane by proteolysis. HBEGF also contains regions rich in basic amino acids, which tether the molecule in the extracellular matrix via association with heparan sulphate proteoglycans. A subgroup of the EGF family of ligands, known variously as 'heregulins' or 'neuregulins', have also been identified. These growth factors are widely expressed in both normal and malignant tissue with a particular prominence in tissues of neural origin. The EGF family therefore represents an example of diversification of biological function through exploitation of a simple structural scaffold.

The fibroblast growth factor family

The fibroblast growth factor (FGF) family was, as the name implies, first identified as mitogenic activities for fibroblast cell lines (such as 3T3) present in ex-

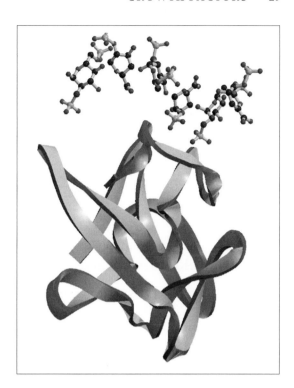

Fig. 2.6 The three-dimensional structure of fibroblast growth factor (FGF) bound to a heparan saccharide. FGFs contain 12 strands linked in trefoil format. Charged sulphate groups on the sugar group interact with basic residues exposed on the FGF surface.

tracts of neural tissues. A defining feature of the FGFs is their strong affinity for the sulphated polysaccharide heparin. This feature proved essential for their ultimate purification and has functional significance in that FGFs always exhibit a paracrine mode of action, since they are tightly bound to extracellular glycosaminoglycans *in vivo* (Fig. 2.6). The original FGFs purified from brain tissue were two closely related molecules, now called FGF1 and FGF2. An intriguing feature of both FGF1 and FGF2 is that both lack secretory signal sequences and, indeed, in most tissues these molecules seem to exhibit a strictly extracellular location. It is clear, however, that some unorthodox mechanisms for export of these molecules exists since, for example, forced expression of FGF2 induces changes in skeletal development, and genetic inactivation of the *FGF2* gene has clearcut effects on blood pressure and neuronal function.

The two prototype FGFs have been joined by a large number of new members of the FGF family. Currently 17 members of the FGF family (FGF1–10 and 15–22) have been identified as functional growth factors, an additional four FGF-related genes have been cloned (termed FGF11–14) and it seems very likely that the current list is incomplete. Moreover, the FGF family seems highly conserved in the animal kingdom; functional FGF homologues have been identified in insects, echinoderms and nematodes. This evidence indicates that, like some other growth factor families, the FGFs have undergone significant expansion and divergence in the course of evolution. This raises the obvious question—what is the biological significance of this rich diversity of growth factor genes? Some insight into this issue comes from studies on the genetics of FGF family members. The generation of mutant mice lacking individual FGF family members clearly points to the fact that each FGF executes a distinct and specific function. For example, animals that lack the *FGF4* gene die at a very early period of development, whereas animals that lack the *FGF5* gene are normal apart from abnormalities in hair growth, which result in angora fur. The most likely explanation is that the expansion and diversification of the FGFs in the course of evolution marches in step with the increased diversification of cell and tissue types of the animal body. The same basic regulatory mechanisms become recruited for different purposes as different cell types and tissue structures arise and 'cross talk' is avoided by diversification of function.

The insulin-like growth factors

There are currently two members of the insulin-like growth factor (IGF) family: IGF-I and IGF-II. As the name implies they are related to insulin in structure and primary sequence but, as might by now be expected, they exhibit quite distinct biological activities. The IGFs are in fact a dramatic manifestation of the central concept that growth factors act to coordinate the size of cell populations and tissues by signalling between cells since their primary biological activity appears to be the control of body mass.

The IGFs were first purified from plasma as agents which mediated the effects of growth hormone on skeletal growth. They are small polypeptides of about 6 kDa which circulate in a latent form associated with a family of specific IGF-binding proteins. Thus, although they are found in body fluids, their activity is almost certainly local in action since they need to be released from binding proteins in order to act. Circulating and local levels of IGF-II are also controlled by the action of a generally expressed cell surface receptor, the type II IGF receptor, whose primary function in mammals is to 'scavenge' IGF-II by importing it into the cell for proteolytic destruction. It would appear therefore that mechanisms exist whose function is to tightly control the amount of IGF-II available to responding tissues.

IGF-I and IGF-II exhibit different patterns of expression: IGF-II is expressed almost ubiquitously during fetal and early postnatal life. IGF-I is primarily expressed by the liver during puberty under the control of growth hormone and levels fall at adulthood. The main action of IGF-I during puberty is on the growth of long bones since the effects of growth hormone deficiency on stature can be experimentally rescued by adminstration of IGF-I.

The primary biological function of IGFs has been revealed by a generation of mice genetically deficient in IGF-II. These mice, although perfectly proportioned, are much smaller than their normal counterparts. Conversely, local over-expression of IGF-II, which occurs, for example, in the case of the human Beckwith–Weidemann syndrome, leads to a disproportionate increase in body mass. This clearly indicates that IGFs act to control the size of tissue populations and, under normal circumstances, achieve this by acting in concert on every cell type in the body. They are, in this sense, universal mediators of cell multiplication. This may explain an additional feature of the IGF axis: IGF-II is subjected to a genetic phenomenon called genomic imprinting. The IGF-II gene is exclusively expressed from the paternal gene during development and the maternally derived gene is silent. Conversely, the type II IGF receptor, whose function is to decrease circulating IGF-II, is subjected to exactly the opposite form of imprinting, whereby only the maternal gene is active and the paternal gene is silent. This may appear at first sight to represent a futile situation since the biological effects of imprinting in the IGF axis would appear to cancel each other out. However, this strange situation almost certainly reflects the central importance of IGF-II levels in the control of body mass in that it reflects an ongoing evolutionary 'battle' between maternal and paternal genomes to control the size of offspring. It has been suggested that, in the context of natural selection, males are best served by having offspring from multiple mothers and would therefore benefit from mechanisms, such as an increase in body mass, which ensure the survival of their offspring. Conversely, mothers have an equal 'interest' in all offspring and carry the major metabolic burden of pregnancy. They would accordingly benefit from a mechanism that decreases the body mass of the fetus. This idea may well explain the existence of elaboration mechanisms to control circulating levels of IGFs *in vivo*.

It may seem surprising that, given the striking role of the IGF axis in the control of tissue mass, the IGFs are, under most circumstances, remarkably feeble inducers of DNA synthesis *in vitro* when compared, for example, with FGFs or PDGFs. However, careful experimental studies of the requirements for cell multiplication *in vitro* (reviewed in Chapter 1) also revealed an almost ubiquitous requirement for IGF signalling in the control of cell multiplication as determined by an increase in population size. The resolution to this paradox almost certainly reflects the action of IGFs in preventing apoptosis rather than inducing cell division (see Chapter 1). This illustrates the important concept of different types of growth factors exhibiting concerted action since, in order for a cell population to increase in size, growth factors (such as the IGFs) which act to prevent cells dying have to act in concert with other growth factors which induce DNA replication and cell division to shape the final tissue size and form.

The transforming growth factor beta family

Thus far we have considered growth factors as exhibiting essentially positive actions: they act to induce events that would not otherwise occur. However, in the same way that cars have brakes and accelerators, the systems that control the multiplication of cell populations have both positive and negative regulators. The transforming growth factor beta (TGFB) family of growth factors, amongst their many biological activities, are striking examples of negative regulators of cell proliferation.

The prototype member of the TGFB family, TGFB1, was first identified as an agent which synergised with the EGF family member TGFA, to induce the transformed phenotype (see Fig. 1.8) in established fibroblast cell lines. The action of TGFB in this assay was, in effect, to modulate (or transform) the activity of another growth factor. TGFB1 is a homodimeric protein composed of two disulphide-linked polypeptide chains of 12.5 kDa which, although exhibiting little primary sequence homology and no common biological function, shares a structural fold with the PDGF family of growth factors. TGFB is expressed in a latent, biologically inactive, complex comprising an inactive pro-form of TGFB bound to a binding protein (the TGFB latency protein) which blocks biological activity. In order for TGFB to be biologically active, at least two separate events are required: proteolytic cleavage of the TGFB precursor and release of the TGF-binding protein (Fig. 2.7). The biochemical mechanisms that bring about acti-

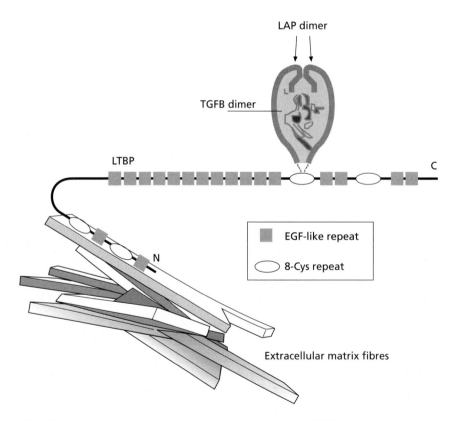

Fig. 2.7 The molecular basis of transforming growth factor beta (TGFB) latency. The TGFB latency protein (LTBP) is tethered to the extracellular matrix via its amino terminus. The TGFB dimer is complexed to the inactive TGFB precursor and the LTBP. (Courtesy of Jorma Keski-Oja, University of Helsinki.)

vation of the latent complex are still poorly understood but the phenomenon of TGFB latency surely indicates that its activity *in vivo* requires specific inhibitory control mechanisms to prevent illicit action.

TGFB1 proves to be the prototype member of a very large and diverse family comprising at least 40 members, many of which have biological functions far removed from the control of cell multiplication. Most closely related in primary sequence to TGFB1 are TGFB2 and TGFB3. These three proteins share many biological functions *in vitro*. Closely related in sequence, but not in function, are the bone morphogenetic proteins (BMPs). These were first isolated from extracts of skeletal tissue as agents which induce ectopic bone formation *in vivo*. It has emerged, however, that the BMP family has many other activities outside skeletogenesis; these include regulation of nervous system development and spermatogenesis.

Another biologically significant set of TGFB-related proteins are the activins and inhibins, which were first isolated as reproductive hormones but are now known to be involved in the development of a variety of tissue systems such as the skin and skeleton. A further TGFB family member with striking biological action is myostatin, which was identified as an inhibitor of muscle growth *in vivo* by regulating the number of muscle cell fibres. Naturally occurring mutations of the myostatin gene have been found in certain breeds of domestic cattle which exhibit a characteristic 'double muscle' phenotype (Fig. 2.8). Mullerian in-

Fig. 2.8 A Belgian Blue bull. This breed has a mutation in the transforming growth factor beta family myostatin gene which produces a characteristic 'double muscle' phenotype. (Courtesy of Tim Bryce and the British Belgian Blue Cattle Society.)

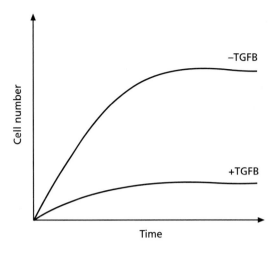

Fig. 2.9 Transforming growth factor beta (TGFB) inhibits the proliferation of epithelial cells in culture. Mouse skin epithelial cells were grown in the presence or absence of TGFB. In the presence of TGFB multiplication of the cells is suppressed.

hibitory substance (MIS) is another TGFB family member whose primary, and so far sole, mode of action is to induce regression of the Mullerian ducts during secondary sexual differentiation of males. As might be surmised from the rich diversity of the TGFB family in mammals, homologues have been identified in a wide variety of species, including, most notably, the *Drosophila* gene decapentaplegic which acts during *Drosophila* development to control dorsal/ventral patterning. As will be seen in later chapters, the existence of evolutionarily conserved growth factor genes permits the powerful tools of fly genetics to be harnessed in unravelling TGFB signalling pathways.

The major action of the prototype TGFBs, TGFB1–3, is inhibition of cell proliferation. They most clearly manifest their activity in the presence of other growth factors. The inhibitory actions of TGFB1 are most strikingly shown in the inhibition of epithelial cell multiplication (Fig. 2.9). Indeed, TGFB1 was purified as an activity found in the culture supernatants of epithelial cells; it prevented their multiplication at high density and therefore, along with depletion of positively acting growth factors such as PDGF, contributed to the control of saturation density (see Chapter 1) of epithelial cells *in vitro*. Careful analysis of the effects of TGFB on cell multiplication shows that its inhibitory action is not directly due to direct inhibition of other growth factors but, in fact, takes place at a point in the G1 phase of the cell cycle that roughly corresponds to the restriction point when cells become committed to complete the cell cycle in response to growth factor signalling. This suggests that TGFB induces a distinct set of cellular processes from positively acting agents such as PDGF, which intersect at the restriction point in the normal cell cycle.

Further biological evidence for an inhibitory function of TGFB *in vivo* comes from genetic studies. Thus, inactivation of the *TGFB1* gene in mice leads to a lethal postnatal phenotype that involves a vigorous autoimmune response. This implies that an important role of TGFB *in vivo* is to inhibit the multiplication of self-reactive immune effector cells. In addition, it might be concluded that loss of TGFB effector pathway function might be associated with hyperplasia and malignancy. This has indeed proved to be the case since loss-of-function mutations in important components of the TGFB effector mechanism have been detected in naturally occurring human tumours.

A second major function of TGFB is the induction of extracellular matrix deposition. Indeed, it is probably this activity which accounts, in part, for its ability to induce anchorage independent growth of established cells in culture in the presence of other growth factors. TGFB may, in this setting, be acting to induce local deposition of extracellular matrix components required for cells to respond to growth factor signalling. The ability of TGFB to elicit extracellular matrix deposition is most strikingly shown by subcutaneous injection of TGFB, which results in a dramatic induction of matrix deposition and consequential fibrosis and scarring. Indeed, it is notable that TGFB expression is a common feature of fibrosis in pathological settings, such as wound scarring, where TGFB is produced by invading macrophages. In fetal wounds, which do not involve macrophage invasion, fibrotic scarring is absent.

Cytokines

Cytokines represent a large multigene family whose diversity of biological functions have evolved by elaboration of a simple structural framework. The name cytokine was originally coined to denote a soluble mediator of immune function. It is now clear that cytokines, as such, have broad-ranging actions outside the immune system and, conversely, some molecules with no overt immune effector function share a common structural fold. The defining feature of the cytokine family of growth factors is a common three-dimensional structure: the four helix bundle. This is a simple generic fold comprising four alpha helices of roughly equal length linked by flexible polypeptide loops (Fig. 2.10). The prototype cytokines are, on this basis, growth hormone and prolactin. Within the four helix layout there are three subgroups. The 'long chain' group have long regions of alpha helix of 20 residues or more. The 'short chain' cytokines, such as interleukin-2 (IL-2) or granulocyte–macrophage colony-stimulating factor (GM-CSF), have regions of alpha helix of typically 10–15 residues and therefore exhibit a more compact form (Fig. 2.10). Finally, there a few examples of dimeric cytokines, such as IL-5, which have eight alpha helices and comprise what is in essence a duplicated version of the four helical form.

Although the cytokine family exhibits many biological functions, it comprises probably the most 'mature' growth factors in terms of their clinical utility. In fact, growth hormone itself was one of the first protein drugs to be administered to human patients and is now routinely used to treat pituitary dwarfism. Erythropoietin is another cytokine that exhibits exquisite target cell specificity: its sole cellular target is the pre-erythrocyte. Erythropoietin can therefore be used to boost circulating erythrocyte levels in individuals suffering from

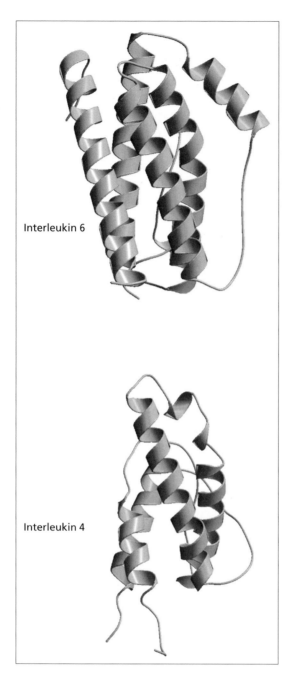

Interleukin 6

Interleukin 4

Fig. 2.10 The cytokine family of growth factors. Shown are the three-dimensional structures of a typical long chain cytokine, interleukin-6, and a short chain cytokine, interleukin-4. Note that both proteins are comprised of a 'core' of four alpha helices but the lengths of the helices are longer in interleukin-6 than interleukin-4. There are also more subtle distinctions between the two proteins, including the presence of an extra short region of helix in interleukin-6 and differences in the stacking angles of the helices.

anaemia. Thrombopoietin is a related cytokine which exerts similar actions on megakaryocyte precursors and can therefore be used to amplify circulating platelet levels. Granloctye colony-stimulating factor (G-CSF) similarly induces the multiplication of granulocytes and can therefore be employed clinically to boost immune effector cell function in individuals receiving chemotherapy.

One reason that these cytokines have rapidly found their way into clinical applications is that they are very specific in terms of the target cell types on which they act. This is due to the fact that, in many cases, they were isolated using assay formats (usually colony formation of specific cell types) specifically designed to identify growth factors which amplified specific haemopoietic target cell populations in the bone marrow. This has led to a detailed description of the growth factor requirements of cells at different stages of haemopoietic development from primitive stem cells to mature differentiated cell types such as the erythrocyte and granulocyte. The details of this provide a splendid example of growth factors acting in hierarchical combinations to elicit expansion and maturation of target cell types. Some haemopoietic cytokines, such as IL-3, seem to act on many cell types, whereas others, such as erythropoietin and G-CSF, are very specific in their action. In addition, as might be anticipated, specificity is often achieved in this system by target cells responding to a combination of growth factors rather than each cell type being defined by a unique growth factor. It is very probable that this general philosophy will be found to apply in other hierarchical developing systems such as the nervous system.

Summary

This chapter has not attempted to document the entire spectrum of growth factor biology and biochemistry but rather to highlight key concepts.

Growth factors come in families; this probably reflects evolutionary diversification of function as the form, composition and physiological roles of animal tissues exhibit increasing sophistication. Growth factors act locally within tissues and a range of biochemical mechanisms exist to regulate their action in time and space. Some growth factors exhibit exquisite cell-type specificity in their action, whereas others appear to be almost universal in their function. Growth factors do not act alone but in teams: their biological actions therefore reflect the context in which they act and this partly explains why the same factor can have different activities in different settings. Above all, however, growth factors are the key extracellular orchestrators of cell multiplication. It is their activity which ultimately controls the size of tissue populations *in vivo*.

Chapter 3: Receptors

Introduction

How can the biological specificity of growth factor function be explained? The answer is that the biological effects are mediated by association with cell surface receptors. Receptors are essentially transducers which have two separate functions; they need to recognise and bind a specific growth factor on the outside of the cell and, once that has occurred, they need to activate a signalling process on the inside of the cell.

The first process dictates the specificity of growth factor action: a cell can only respond to a particular growth factor if it expresses the appropriate complementary receptor. The second process dictates the functional response of the cell to receptor engagement since that determines the type of signalling processes that are activated. Although these two functions are clearly distinct, the second is dependent upon the first. There must therefore exist a means by which ligand engagement on the outside of the cell is translated into activation of biochemical signalling processes on the inside of the cell. In this respect transmembrane receptors are often likened to enzymes—the 'catalytic' domain of the receptor elicits intracellular signalling and the 'regulatory' domain restrains the activity of the catalytic domain until a ligand is engaged.

Growth factor receptors have four basic configurations which we will deal with in turn. The first are quite literally enzymes, since they are characterised by having intrinsic tyrosine kinase activity. The second are defined by structural features of the ligand recognition domain: these are the so-called cytokine receptor family. The third mediate the biological responses of the TGFB family of growth factors and are characterised by having an intrinsic serine/threonine kinase enzymatic activity in the intracellular domain. The final family are members of the larger 'seven transmembrane' family of receptors, also employed by peptide hormones and neurotransmitters; these will be considered in Chapter 4.

Upon ligand binding these basic receptor forms generate three classes of biochemical signals: tyrosine phosphorylation, serine/threonine phosphorylation and membrane phospholipid hydrolysis. Each type of receptor can, directly or indirectly, generate multiple types of signals. An understanding of how these signals are generated as a result of ligand binding comes from understanding receptor structure.

Intrinsic tyrosine kinase receptors

The intrinsic receptor tyrosine kinase (RTK) family of receptors is defined by the presence, in the cytoplasmic domain, of regions of amino acid sequence conservation which corresponds to a protein kinase with intrinsic tyrosine kinase activity. The RTK family is large and includes the receptors that mediate signalling by many of the growth factors discussed in Chapter 2, including PDGF, FGF and EGF. The vast majority of this family of receptors have a single transmembrane spanning region (the exception are the receptors for IGFs and insulin which are expressed as disulphide-linked receptor dimers). This feature immediately raises the question as to how a signal can be transmitted across a membrane by means of a single transmembrane domain.

A key feature of the RTK family of receptors is that, very frequently, an individual growth factor binds to multiple receptors. Thus, for example, the FGF family of ligands interact with a family of four FGF receptors, FGFR1–4, and the PDGF family interact with a pair of closely related receptors, PDGFRa and PDGFRb. At least part of the reason for multiple receptors is that each member of a receptor family exhibits differences in either the specificity of ligand recognition or signal transduction (or both). The specificity of ligand recognition in this case is rarely absolute but, more frequently, combinatorial in nature. Thus, the PDGFRa is able to bind both PDGFA and PDGFB, whereas the PDGFRb is only able to bind molecules containing a PDGFB chain. This means that the three different isomers of PDGF, AA, BB and AB, each exhibit a unique combination of receptor targets. The same is true for the FGFR family where individual ligands bind to characteristic combinations of target receptors. FGF1, for example, is a 'universal' ligand, able to interact with all four receptors, whereas FGF9 is much more severely restricted in scope, being only able to bind FGFR3 with high affinity. This theme is further exemplified in the FGFR family where specific alternative splicing events in the extracellular domain of FGFR1–3 exert an additional powerful influence on ligand recognition.

This idea of the diversity of receptors having a combinatorial role leads to the proposal that each receptor should mediate a specific range of biological functions. This is borne out by the results of genetically inactivating RTK receptors in mice. Thus, mice which are deficient in FGFR2 die very early in development (with a remarkable resemblance to mutations in the ligand FGF4), whereas mice deficient in FGFR3 show enhanced growth of the long bones. As more of this type of genetic evidence emerges, it is beginning to seem clear that the apparent diversity of receptor forms is a mechanism for achieving exquisite specificity in biological function.

The RTK family of receptors exhibit considerable heterogeneity of form in the ligand recognition domain (Fig. 3.1). Thus, the PDGF and FGF receptors have an extracellular region that is principally composed of multiple domains related in structure to the antigen-combining domains of antibodies (Ig domains). The EGF receptors, on the other hand, are composed of multiple blocks of cysteine-rich regions whose three-dimensional structure is hard to predict. As more RTK class receptors have been cloned and sequenced, it emerges that their extracellular domains are generally composed of a melange of a limited

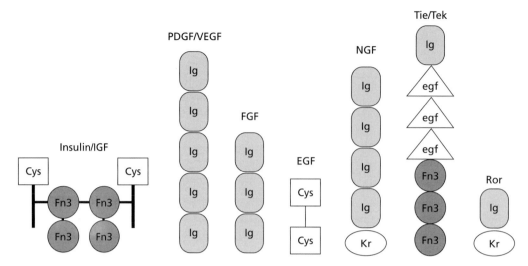

Fig. 3.1 The modular structure of tyrosine kinase receptor (RTK) extracellular domains. The extracelluar regions of RTKs are diverse in form and composed of differing types of protein modules linked in assorted arrangements (Cys, cysteine; egf, egf module; Fn3, fibronectin type III module; Ig, immunoglobulin; Kr, kringle). Shown here are a sample of different RTK family receptors. Some protein modules, such as the Ig domains, are found in many receptors whereas others, such as the kringle (Kr) domain, are found in relatively few.

number of structural motifs arranged in different fashions (Fig. 3.1). Ligand recognition by the RTK family therefore seems to involve a diversity of strategies based upon 'mix and match' of a limited number of structural motifs.

This rich diversity of structural strategies for ligand recognition may be contrasted with the cytoplasmic domain, which is much more highly conserved amongst RTK family members. Indeed, the strong sequence conservation in the RTK family kinase domain has proved to be a powerful lever in the identification of novel receptors since they can be cloned on the basis of sequence conservation rather than function. This difference in diversity between the outside and inside of the cell suggests that each region of RTK receptors has evolved separately. This is strongly supported by the very useful finding that it is possible to experimentally swap the extracellular domains of different receptors whilst retaining normal signalling function. This also suggests that activation of receptor signalling does not require any specific features of the extracellular domain other than its ability to bind ligand with high affinity. This aspect provides an important clue as to how receptor signalling might be activated.

The principal feature of the cytoplasmic domain of RTK receptors is the enzymatic domain encoding the intrinsic tyrosine kinase activity. This region, based upon sequence comparisons, comes in two types. The first, exemplified by the EGF receptor, is where the amino acid sequences characteristic of the enzymatic domain are found in a contiguous block. The second, more common,

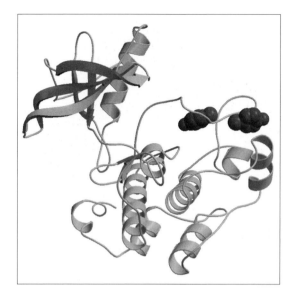

Fig. 3.2 The X-ray structure of the fibroblast growth factor receptor (FGFR) tyrosine kinase domain. This is composed of two regions. The N-terminal region, containing a single alpha helix and multiple beta strands, binds the ATP substrate. The C-terminal region, composed of regions of alpha helix, binds the peptide substrate. An important feature of the C-terminal domain is the 'activation loop' containing two prominent tyrosine residues (shown in CPK format) which blocks access of the substrate to the binding site.

type, such as the FGFR family, is where the linear amino acid sequence is broken by a region of varying lengths of 'insert' sequence. An important feature of the 'insert domains' is that they invariably contain multiple tyrosine residues which act as substrates for activated receptor kinases.

The catalytic function of the tyrosine kinase domain has been powerfully informed by determination of the three-dimensional structure of the FGFR1 and insulin kinase domains. This reveals (Fig. 3.2) a typical bilobed kinase motif. The adenosine triphosphate (ATP) substrate binds to the N-terminal lobe and the peptide substrate binds to the C-terminal lobe. Obviously, in order for the peptide substrate to become phosphorylated, a conformational change must occur which brings the two lobes into close physical proximity. In this respect the most important feature of the structure is the presence of the 'activation loop', a region of the polypeptide chain that forms a loop which sits between the two lobes blocking access of bound polypeptide substrate to ATP. In most cases the predicted sequence corresponding to the activation loop contains one or more tyrosine residues which are phosphorylated upon receptor activation.

This raises an interesting question. In order for kinase activation to occur, the activation loop must be moved away from the substrate binding region. There must, therefore, occur a conformational change in the kinase domain brought about, directly or indirectly, as a consequence of ligand binding to the extracellular domain. This is achieved by the ligand acting to dimerise receptors, thereby bringing two receptor kinase domains into close physical proximity. This is the key to understanding the mechanism of receptor activation by polypeptide growth factors.

The best evidence for the significance of receptor dimerisation in growth fac-

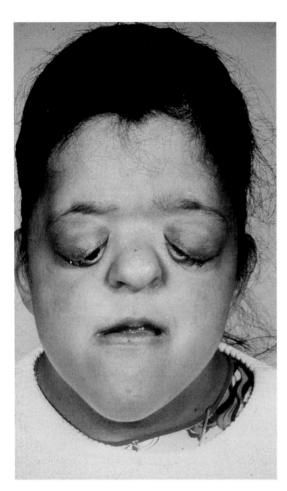

Fig. 3.3 Crouzon syndrome arises as a result of mutations which introduce unpaired cysteine residues into the extracellular region of FGFR2. This leads to activation of receptor signalling in the absence of ligand, producing premature fusion of the skull sutures and the resulting craniofacial phenotype.

tor signalling comes from the study of different types of receptor mutants. It is intuitively obvious that, if dimerisation is the key to ligand activation, mutations which induce dimerisation without receptor binding should activate receptor kinase activity in the absence of the growth factor ligand. This is indeed the case. Various types of naturally occurring, or manmade, mutations in RTK receptors result in activation of kinase activity. A very good example is the inherited congenital craniofacial malformation Crouzon syndrome (Fig. 3.3), caused by activation of FGFR2 signalling in the developing skull. This is brought about by the presence of mutations in the extracellular domain which introduce free cysteine residues. As a consequence, mutant receptors are free to undergo disulphide-mediated dimerisation with each other and this results in activation of receptor kinase activity and the resulting dysmorphology. Similar types of mutations have been found in other types of RTK receptors and these result in transformation of the mutant cells and tumour formation. This

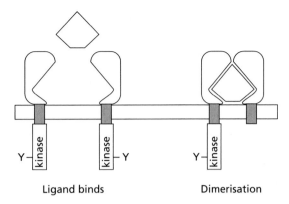

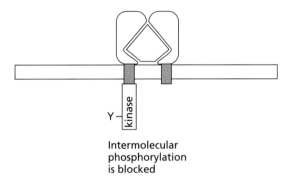

Fig. 3.4 Dominant negative mutant receptors lack a functional kinase domain. Receptor dimers are formed in the presence of ligand between mutant and wild-type receptors. These receptor complexes are unable to signal, showing that intermolecular dimerisation of kinase domains is required for signalling.

evidence clearly shows that dimerisation is sufficient to activate tyrosine kinase activity.

It is possible to approach the question of receptor dimerisation from a different angle. If the point of dimerisation is to bring two cytoplasmic domains into close physical proximity, what would happen if one of the kinase domains is missing? If kinase activation did not occur it would prove that the purpose of dimerisation is to bring the kinase domains together. This is indeed the case. Deletion of the cytoplasmic domain of all RTKs tested produces 'dominant negative' mutant receptors. These mutants have the ability to block the signalling functions of their normal counterparts (Fig. 3.4). Expression of dominant negative forms of receptors *in vitro*, and *in vivo*, can clearly inhibit ligand-mediated signalling in many experimental situations. The fact that this approach works shows that the purpose of ligand binding is to dimerise receptors.

The action of dominant negative receptor mutants not only provides strong evidence for the biochemical features of receptor signalling but also provides a powerful technique for dissecting growth factor function *in vivo*, since it is implicit in this approach that the action of dominant negative receptor mutants will only be revealed in the presence of ligand. The combined evidence from these two types of receptor mutants is very clear: receptors elicit biological effects as a result of the activation kinase activity. This naturally raises a series of interrelated questions.

How do growth factor ligands cause receptor dimerisation? Several schemes can be envisaged. These include the induction of conformational changes in the extracellular domains such that receptors undergo intermolecular association. A related idea is that the function of unbound extracellular domains is to prevent intermolecular association and this essentially inhibitory function is relieved upon ligand engagement. The most probable explanation is, unfortunately, not only more prosaic but also more biologically plausible. That is that growth factors have biochemical features that allow them to bind two receptors simultaneously. This is fairly obvious in the case of dimeric ligands such as the PDGF family (Fig. 3.5) but also turns out to be true for ligands such as EGFs and FGFs. Analysis of the composition of receptor/ligand complexes reveals the existence of two molecules of ligand and two molecules of receptor: ligand dimerisation occurs in this case by non-covalent means.

How does dimerisation of cytoplasmic domains elicit activation of tyrosine kinase activity? The answer to this question is by no means clear. Consideration of the structure of tyrosine kinase domains (Fig. 3.3) suggests that, whatever happens, the consequence must be to cause displacement of the activation loop away from the substrate-binding region to permit access to ATP bound to the N-terminal lobe. Two general theories can be put forward. This first is the 'crowded room' model. This suggests that the essential consequence of dimerisation is

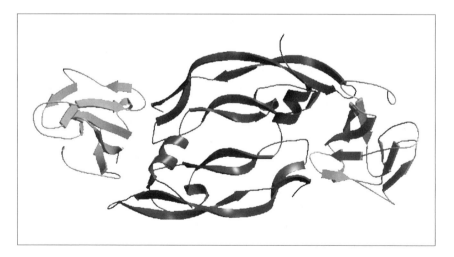

Fig. 3.5 The three-dimensional structure of vascular endothelial cell growth factor (VEGF) bound to receptor. VEGF is composed of two anti-parallel dimers, each of which contacts a single molecule of receptor. As a result, two molecules of receptor are bound by one molecule of ligand.

to produce a very high local concentration of substrates by virtue of the propinquity of the two ATP-bound forms of the kinase domain. Given an intrinsic mobility of the activation loop it follows that, sooner or later, substrate will be phosphorylated. The substrate in this case is the activation loop of one of the partner receptors in the complex. The consequence of activation loop phosphorylation would be to move it into an 'active' conformation, thereby permitting activation of the partner kinase. This model places great importance on the significance on tyrosine phosphorylated forms of the activation loop and is supported by the fact that mutation of tyrosine residues in the activation domain diminishes receptor kinase activity.

The second, 'nudge in the elbow', model is more subtle but also in keeping with generic features of protein kinases. This suggests that the two kinase domains approach each other in some specific orientation which, upon contact, causes a simultaneous conformational change in both partners such that the two activation loops are moved away from the substrate-binding site of one receptor into the, by now, active site of the other partner. This model is also consistent with available data but is currently very difficult to test experimentally. A key feature is the idea that the structural relationship between the extracellular domain and the cytoplasmic domain is somehow preserved, despite the apparent ease with which the former can be interchanged whilst preserving signalling function.

The most important question, however, is the simplest. How does activation of tyrosine kinase activity lead to induction of cell proliferation? The basic answer is that activation of kinase activity has two consequences. The first, which might be predicted from generic features of protein kinase regulated pathways in other systems, is that phosphorylation of substrate proteins alters their biochemical activity. The second, and undoubtedly the more generally important, is that it induces the relocalisation of proteins within the cell.

There is considerable experimental evidence to show that the acquisition of tyrosine kinase activity upon ligand engagement is vital to the propagation of mitogenic signals. The basic experimental design, which can be subject to many elaborations, is to create mutant receptors which either lack kinase activity or lack tyrosine residues which form the immediate substrates for the kinase. The essential result in both categories of experiments is clear: the most obvious substrate for an activated receptor kinase is another receptor and receptors need to be phosphorylated for signalling to occur.

Deletion of key residues required for kinase function, such as those involved in binding the ATP substrate, result in the creation of receptors that are inactive in terms of signalling functions and exhibit dominant negative properties. This class of experiment clearly shows that kinase activity is key to RTK function. It should, however, be noted that, in reality, the consequences of this type of mutation may be more complex, in that it appears that tyrosine kinase activity may well play an important role in the control of receptor trafficking and thereby, indirectly, the number of receptors on the cell surface.

The second class of mutants is equally informative and leads, from the classical perspective on kinase function, to an unexpected conclusion. A variety of experimental approaches have shown that, upon activation of receptor kinase activity a multiprotein complex is formed with the activated receptor. Examination of the composition of these complexes, combined with mutagenesis studies, reveals that the specific proteins associate with phosphorylated receptors and this is mediated by phosphorylated tyrosine residues. It seems, therefore, that the function of the active kinase is to create, by phosphorylation, docking sites for specific kinase-binding proteins. The primary mechanism of signal transduction therefore involves relocation of proteins within the cell.

This raises two questions: what is the identity of these substrate proteins and how do they recognise phosphorylated receptor? The two questions turn out to be linked in that the association of substrates with phosphorylated receptors occurs by means of families of specific protein motifs whose function is to bind peptide sequences containing phosphorylated tyrosine residues. These families, whilst sharing a common structure, exhibit specificity in terms of the local amino acid sequence surrounding the target tyrosine residue. The identity of proteins which bind to specific activated receptor is therefore dictated by the amino acid sequence of the docking site in the target receptor. By this means it is possible for related receptors which bind the same ligand to elicit different signals by means of different combinations of receptor-associated proteins. The use of specific protein motifs to recognise tyrosine phosphate docking sites also means that putative substrate proteins can be identified by virtue of the presence of the appropriate protein motifs.

The first such recognition motif to be identified was the SH2 domain. The three-dimensional structure of the SH2 domain (Fig. 3.6) shows how recognition of the tyrosine phosphate is achieved. The SH2 domain can be roughly viewed as resembling a cupped hand with a binding site for the tyrosine phosphate in the 'palm' region. The affinity of the SH2 domain for a specific polypeptide sequence is dictated by the identity of the amino acids surrounding the tyrosine phosphate binding site. Another class of docking motif found in RTK substrates is the phosphotyrosine binding (PTB) domain (Fig. 3.6). This is formed from two alpha helices which flank a beta-stranded region forming a 'cleft', into which the target peptide sequence fits. Different RTK protein substrates have different combinations of SH2 and PTB domains with differing specificities for target peptides containing a phosphorylated tyrosine residue. The repertoire of proteins which complex with an activated receptor will therefore be defined by the exact sequence of the docking sites in the receptor itself (Fig. 3.7).

The intrinsic tyrosine kinase family of receptors has introduced three important ideas: the function of ligand is to oligomerise receptors; receptors have a modular construction (in particular the 'inside' and 'outside' regions seem to be interchangeable); and, finally, the molecular basis of signalling has as a central feature the induced formation of multiprotein complexes.

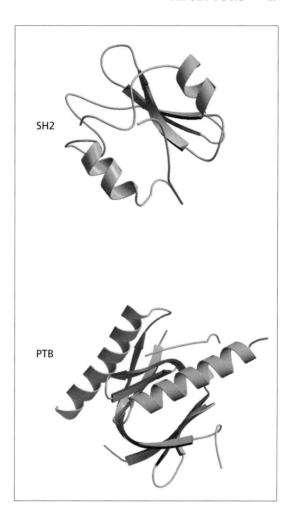

Fig. 3.6 The three-dimensional structure of the SH2 domain and the PTB domain. Each domain binds peptides containing phosphorylated tyrosine residues in a sequence-specific fashion.

The cytokine receptor family

The cytokine receptors are defined by virtue of structural features of the extracellular domain. This is almost certainly because the cytokine family of receptors interacts with ligands with a shared structural fold: the four helix bundle family described in Chapter 2. The cytokine family therefore transduces the signals of a variety of therapeutically important ligands such as Epo, G-CSF and growth hormone.

Sequence alignment of the extracellular domain of the cytokine family of receptors reveals the existence of a signature sequence motif of about 200 amino acids: the 'cytokine homology domain' (CHD). The most obvious defining feature of the CHD is the presence of a short sequence at the C-terminus containing the conserved sequence Trp,Ser, X, Trp, Ser—the so-called WSXWS

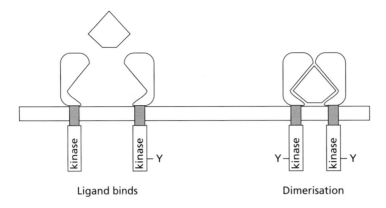

Ligand binds   Dimerisation

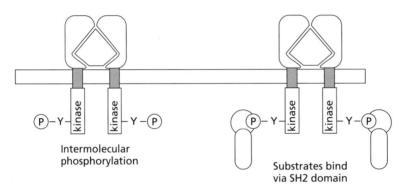

Intermolecular
phosphorylation   Substrates bind
                             via SH2 domain

Fig. 3.7 Summary of receptor tyrosine kinase (RTK) signalling. The ligand binds to the extracelluar domain, inducing receptor dimerisation. This results in intermolecular phosphorylation of receptor cytoplasmic domains on tyrosine residues. Phosphorylated tyrosine residues act as docking sites for protein substrates containing specific protein-recognition modules.

box. Closer inspection also reveals a number of other sequence conservations, in particular a characteristic group of cysteines in the N-terminal portion and a proline-rich 'hinge region' located in the centre of the CHD.

The meaning of this sequence conservation becomes clear in the context of the three-dimensional structure of the CHD. A number of CHD structures are available, including the structure of the growth hormone receptor complexed with growth hormone ligand (Fig. 3.8). These are very revealing since they also directly describe the mechanism by which ligand activates receptor. This shows that the two CHDs bind a single molecule of ligand via two discrete sites on the ligand. The mechanism of activation is therefore, like the RTK family, receptor dimerisation.

The isolation and characterisation of more members of the cytokine family

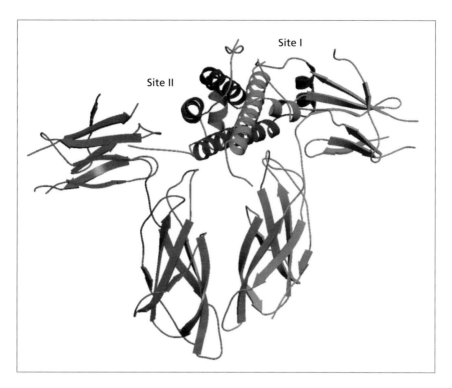

Fig. 3.8 The three-dimensional structure of the cytokine growth hormone complexed to receptor. Two molecules of receptor bind to different recognition sites on the ligand. The cytokine homology domain of the receptor is composed of two seven-stranded modules linked at approximate right angles. This reveals residues on inter-strand loops which contact ligand.

of receptors reveals additional levels of complexity beyond the 'simple' example of growth hormone. First, many cytokine receptors have additional sequence motifs in their extracellular region. Most commonly these are immunoglobulin-like domains and in some cases so-called fibronectin type III repeats (FNIII, named after the prototype fold of the cell-adhesion protein fibronectin). The increase in structural complexity is paralleled by an increasing sophistication in mode of action. Some cytokine-type receptors, such as gp130, are able to interact with multiple ligands and, depending upon the ligand, form heteromeric complexes with several other types of receptor. In these systems there are so-called common subunits (which bind multiple ligands) and specific subunits (which bind a single ligand). A further feature is that many specific subunits act in secreted or membrane-tethered forms. Their function is purely to bind ligand and they cannot, by definition, participate directly in the process of signal transduction. Examples include the specific receptors for IL-6 and ciliary neurotrophic factor. These soluble receptors are, however, important for ligand specificity since they are required for the formation of a higher order complex with receptors that do have signalling functions. Thus, in the case of IL-6 the signalling receptor gp130 is dimerised by the formation of a hexameric complex with two molecules of IL-6 and two molecules of the soluble IL-6 receptor (Fig. 3.9). The cytokine family of receptors conform to the basic concept of receptor oligomerisation as the central mechanism of activation but achieve this by sophisticated use of combinations of receptors and ligands with different affinities. The outcome in all cases is a dimer of two transmembrane receptors which execute signalling functions. Depending upon the precise combination of receptors formed in the complex, different pairs of intracellular domains are brought together (Fig. 3.9).

The intracellular domains of cytokine receptors at first sight offer few clues as to the identity of their signalling mechanisms. They are generally relatively short compared with the RTK family of receptors and have no sequence elements indicative of any intrinsic enzymatic activity. In fact, the cytokine family of receptors signal via a mechanism that is very similar to that of the RTK family—tyrosine phosphorylation. This is because their cytoplasmic domains form a complex with a specific family of soluble tyrosine kinases, the JAK (Janus or 'just another kinase'). family. The JAK family currently comprises three members: JAK1, JAK2 and TYK2. The JAK kinases have two regions of homology to classical tyrosine kinases but only the C-terminal kinase domain is biochemically active. Mutation studies show that specific JAK kinases associate with specific sites in the receptor cytoplasmic domains and that different receptors associate with different JAKs. This means that different combinations of JAKs can be brought together by dimerising different pairs of receptors. This allows different types of signals to be generated by different combinations of receptor (JAK) dimers (Fig. 3.10).

Given the use of JAK kinases for receptor signalling it will be not be a surprise to learn that activation of cytokine-type receptors by ligand leads to activation of JAK kinase activity and concomitant phosphorylation of both receptors and kinases forming docking sites for substrate proteins containing appropriate phosphotyrosine binding motifs. In this respect the types of signals produced by cytokine receptors are similar to those of the RTK class. A striking difference comes, however, from the finding that another specific type of substrate is also bound to the cytoplasmic domain of cytokine receptors. These are the STATs (signal transducers and activators of transcription), whose function (described further in Chapter 5) is to activate gene transcription. STATs are phosphorylated by JAKs and dissociate from the receptor complex and relocate to the nucleus (Fig. 3.10). This is an example of the importance of protein relocation in mitogenic signalling in reverse—in this case it is the breakdown of a 'latent' signalling complex that leads to activation of signalling mechanisms.

There are multiple related STAT proteins that can, in addition, be expressed in multiple forms by alternative splicing. Each STAT family member exhibits a specific repertoire of cytokine receptor associations, so different pairings of STATs can be achieved by different pairings of receptors. Added to combinator-

Fig. 3.9 Different cytokine signalling complexes can be formed from common elements. The gp130 cytokines contain three receptor-recognition sites. Some receptors bind a single site and others bind two. As a result, complexes with different components and stoichiometries can be formed containing common elements. (From Bravo, J. & Heath, J. (2000) *The EMBO Journal* **19**, 2399–2411, reproduced with permission of Oxford University Press.)

ial mechanisms of receptor dimerisation and JAK kinase association, it can be immediately seen how, by 'shuffling the pack', an enormous diversity of signals can be generated from a simple set of related proteins.

The serine/threonine kinase family of receptors

Receptors that mediate signalling via the transforming beta superfamily of ligands (see Chapter 2) are, at first sight, biochemically quite different from the

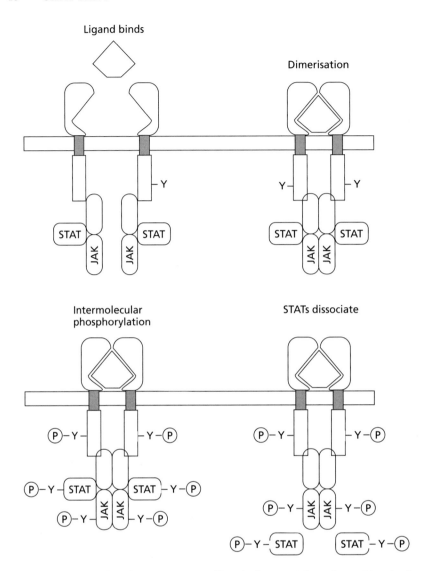

Fig. 3.10 Summary of cytokine receptor signalling via the JAK/STAT pathway. Dimerisation of the receptors by ligands occurs. This induces intermolecular phosphorylation of JAK kinases associated with the cytoplasmic domains. Activated JAK kinases phosphorylate STAT substrates which dissociate from the receptor complex.

RTK and cytokine family of receptors and yet it will become clear that their mechanism of activation is actually very similar.

Of the Ser/Thr family of receptors defined, there are currently 14 members which exhibit a common canonical structure (Fig. 3.11). They have relatively short extracellular domains and relatively little is known about the structural details of receptor recognition by the ligand. There is a single transmembrane

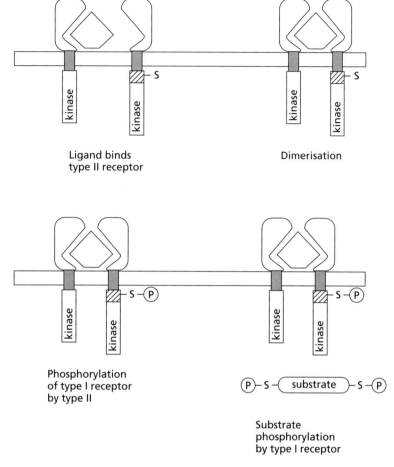

Ligand binds
type II receptor

Dimerisation

Phosphorylation
of type I receptor
by type II

Substrate
phosphorylation
by type I receptor

Fig. 3.11 Summary of receptor signalling via transforming growth factor beta family receptors. Dimerisation of the receptor by ligands occurs. This results in phosphorylation of the type I receptor by the type II receptor kinase. The activated type I receptor phosphorylates receptor substrates such as SMADs, which dissociate from the complex.

domain and a cytoplasmic domain which is almost entirely composed of the protein kinase domain with intrinsic serine threonine kinase activity.

This design, combined with the dimeric nature of TGFB type ligands (Chapter 2), immediately suggests that the basic mechanism of kinase activation must be ligand-induced receptor oligomerisation and, indeed, the same type of evidence (e.g. dominant negative receptor mutants) as produced for RTK-type receptors (see above) shows that, in general terms, this is the case. However, the use of oligomerisation to achieve receptor activation by the Ser/Thr family displays some interesting subtleties.

Sequence characterisation of the cytoplasmic domains of the Ser/Thr family of receptors shows that they can be grouped in two classes — type I and type II —

which are related to each other by about 40% identity. A striking feature of the type I cytoplasmic domains is the presence of a short 30 amino acid sequence in the juxtamembrane region N terminal to the kinase domain. This is the so-called GS domain which reflects the presence of multiple glycine and serine residues, the latter being candidate substrates for phosphorylation. Analysis of the ligand specificity of the Ser/Thr receptors reveals an additional intriguing distinction between type I and type II receptors. Type I receptors bind ligands with relatively low affinity, whereas type II receptors bind specific ligands with high affinity. This is reinforced by analysis of dominant negative type I mutant receptors that are able to block signalling via multiple related ligands. This pattern is highly reminiscent of the common and specific subunits of the cytokine family of receptors. Indeed, following this analogy it appears that, for specific TGFB subfamilies, there exist subfamily type I receptors (for example, for the activins, TGFBs and BMPs) and highly specific type II receptors.

A third distinction between type I and type II receptors lies in the activity of the kinase domain. Type II receptors exhibit constitutively active kinase activity, whereas type I receptor kinases, like the RTK and cytokine family, require receptor oligomerisation to be active. Which receptor activates the signalling pathways? Logic suggests it must be the type I receptor and, indeed, activating mutants of type I receptors have been created in which the kinase is constitutively activated. These mutant receptors signal in the absence of association with either ligand or the type II receptor. Conversely, mutations which eliminate type I kinase activity block the ability of TGFB to signal, although they retain the ability to bind TGFB and the type II receptor. Clearly, in order for TGFB to elicit a biological effect the kinase of the type I receptor must be activated and this is associated with the formation of a complex with the type II receptor.

How is type I receptor activation brought about? The answer lies in the GS domain. The formation of a ligand-mediated complex with the type II receptor leads to phosphorylation of the type I receptor on serine residues in the GS domain. GS domain phosphorylation is required for type I kinase activation since mutation of the target serine residues abrogates type I receptor function. What phosphorylates the type I receptor GS domain? The answer is the constitutively active kinase of the type II receptor, which is brought into close physical proximity by the formation of the receptor complex (Fig. 3.11). In this respect it is convenient to view the type I receptor as the principal substrate for the type II receptor. The TGFB family therefore exhibits an intriguing variation on the theme of receptor oligomerisation in that only one partner in the complex, the type I receptor, executes intracellular signalling functions.

If the Ser/Thr kinase family of receptors are indeed fairly typical receptors, we should expect to find that receptor kinase activation will lead to the formation or destruction of some form of multiprotein complex associated with activated receptors. In fact, it has so far proved difficult to isolate proteins that can associate with the cytoplasmic domains of Ser/Thr receptors which have any proven involvement in the biological responses to receptor activation.

The identification of key receptor substrates has come from exploitation of the conservation of the TGFB pathway in animals with accessible genetics, such as *Drosophila* and *Caenorhabditis*. A number of genetic strategies have revealed a common, conserved family of proteins which appear to be essential for TGFB superfamily signalling. These are the SMAD family of transcription factors. (The name is a fusion between the *Caenorhabditis* homologue sma-1 and the *Drosophila* gene mothers against decapentaplegic MAD). The SMAD family in mammals currently comprises at least nine members. Genetic studies in both flies and mice clearly indicate that SMAD action accounts for most, if not all, of the biological functions of the TGFB superfamily of ligands.

SMADs have a common structure, being composed of two domains linked by what is predicted to be a flexible linker region. The C-terminal domain appears essential for effector function and activation of gene transcription, whereas the N-terminal domain exerts an inhibitory effect on SMAD function. How are SMADs activated by Ser/Thr kinase receptors? They are directly phosphorylated by activated type I receptors in the N-terminal region, presumably following some kind of transient association with the activated receptor complex. This results in relocation of the phosphorylated SMAD to the nucleus, protein dimerisation, relief of the inhibitory functions of the N-terminal domain association with DNA and activation of gene transcription. In this respect there are strong parallels between the SMAD family and the STAT family in terms of their mechanism of activation by signalling receptors.

Summary

Ligands exert their biological effects by association with specific receptors. The portion of the receptor on the outside of the cell dictates association with ligand and the portion on the inside dictates the type of signals that are generated. A common strategy is that ligands induce the formation of a multiprotein complex which initiates the process of signal transduction. The formation of a complex of intracellular domains leads to activation of kinase activity, either intrinsic to the receptor or non-covalently associated. A consequence of kinase activation is the relocation of proteins within the cells, either to form multiprotein complexes with the receptor or, conversely, for relocation of receptor-associated proteins into the nucleus. A central theme is, therefore, the importance of altering protein location in the induction of mitogenic signalling. Finally, at all levels of the process from ligands and receptors through to effectors, we see evidence of combinatorial mechanisms at work—families of related proteins can be put together in different combination to allow a wide diversity of structures to be created from a small number of component parts.

Chapter 4: Intracellular Signals

Introduction

The ligand-mediated formation of a receptor complex provides the trigger for relocalisation of proteins within the cell to form a signalling complex. In this chapter we will consider the components of this complex and the biochemical pathways with which they engage. It is useful to consider the intracellular domain of, say, an intrinsic tyrosine kinase receptor as being essentially a 'coding' device or a blueprint for building a protein signalling machine. The 'code' on this case is the precise disposition and combination of docking sites for signalling components which are activated in response to receptor oligomerisation. Thus, part of the significance of there being multiple receptors for individual ligands is that the same ligand can activate different combinations of signals by interaction with receptors bearing different intracellular 'codes'. The coding idea can be extended a step further. If different types of cells had different combinations of signalling components, the one receptor could elicit different signals in different types of cells by virtue of the cell type specific combination of signalling components that are assembled. This is precisely the reason why growth factor receptors can induce cell multiplication in one cell type and cell differentiation in another.

This represents a very elegant mechanism for extracting the maximum use out of a limited number of components and is yet another manifestation of the ubiquitous application of combinatorial logic to biological signalling processes. It is also a recipe for fearsome, and probably unmanageable, complexity. In this light there is no single answer to the question 'what signals are generated by this receptor?'—in fact, it might be easier to define what signalling cannot be employed by a particular receptor. However, it is possible to define a number of defined channels or biochemical pathways down which intracellular signals are passed in the process of activating cell proliferation. From this it emerges that there are a limited number of principles involved. All signalling pathways revolve around modulation of the activity of substrate-specific protein kinases. All signalling pathways involve the formation and destruction of protein-signalling complexes that are mediated by protein 'switches'. The activity of these switches frequently exploits, directly or indirectly, the lipid components of the plasma membrane. Our analysis of receptor-activated cellular signalling will proceed with characterisation of the principal signalling channels.

Lipid-mediated signalling pathways

Aside from the receptor classes described in Chapter 3, an additional category of

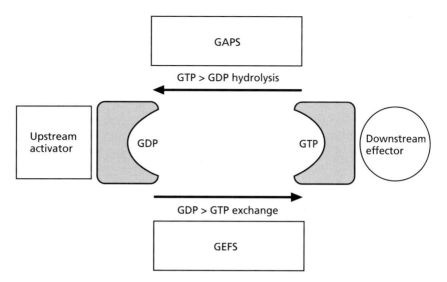

Fig. 4.1 The action of G proteins. In the GTP bound state the G protein interacts with, and activates, downstream effector proteins. The G protein has intrinsic GTPase activity which results in the hydrolysis of GTP to GDP, uncoupling the G protein from its effectors. GDP is exchanged for GTP upon interaction with an upstream activator, such as an active receptor complex. The rate of exchange between GTP and GDP is also controlled by associated proteins: exchange factors (GEFs) promote GDP to GTP exchange and GTPase activating proteins (GAPs) promote GTP to GDP exchange.

receptors has been implicated in mitogenic signalling. These are receptors of the classical seven transmembrane spanning type, more frequently associated with the activity of neuropeptides and hormones. However, some ligands which act via these receptors have been found to induce progress through the cell cycle in certain types of cells. Examples include the neuropeptides bombesin and bradykinin, the protease thrombin, and a very large class of polypeptide mediators of inflammation: the chemokines.

The signalling mechanisms used by these receptors are relatively simple in that they predominantly involve the use of a special type of molecular 'switch' called a G protein. A G protein can be considered to resemble a switch in that its activity is regulated by the binding and hydrolysis of guanosine triphosphate (GTP) (Fig. 4.1). A typical G protein binds GTP as a co-factor but, at the same time, exhibits intrinsic GTPase activity so that, under normal conditions, occupancy by GTP is transient since it is hydrolysed back to GDP. A G protein has a second critical design feature: it is able to interact with different protein partners depending upon whether it is occupied with GTP or GDP. Association with 'upstream' partners, in this case ligand-occupied receptors, induces the G protein to exchange GDP for GTP; this induces the G protein to dissociate from the receptor and associate with the 'downstream' partners. This association is terminated by the hydrolysis of the GTP to GDP, causing dissociation from the downstream partners and reassociation with the upstream partners, bringing us back to where we started. This GDP/GTP switching mechanism therefore dictates which set of protein partners is engaged. In addition, this system is intrinsically transient in that duration of downstream partner association is controlled by the rate of GTP hydrolysis. This means that a G-protein-mediated process is rapidly terminated if the upstream partner can no longer engage.

What type of mitogenic signalling processes are linked into G-protein-linked switches? These signalling processes must involve a downstream partner protein whose activity is regulated by the GTP-bound G protein. An important set of enzymes that are linked into G-protein-coupled receptor systems are phospholipases; these are involved in the hydrolysis of plasma membrane lipids exposed on the cytoplasmic face of the membrane. In particular, activation of receptor signalling by the small peptide mitogens, such as bombesin or chemokines, induces the activation of phospholipases which degrade a specific class of plasma membrane lipids containing inositol polyphosphate headgroups (Fig. 4.2). The action of these enzymes produces two sets of molecules: water-

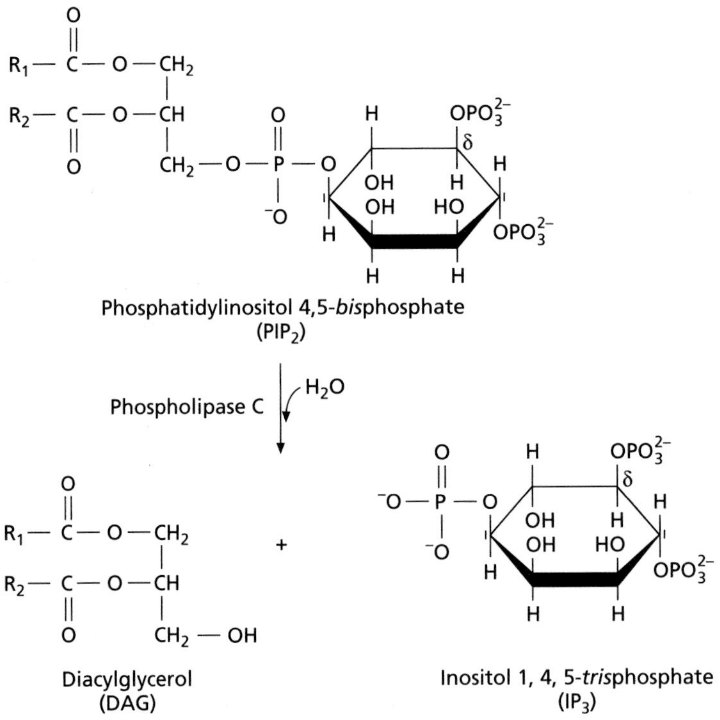

Fig. 4.2 Three-dimensional structure of PIP_2 and the products of hydrolysis of phospholipase C: IP_3 and DAG. IP_3 is water soluble and binds to receptors in the endoplasmic reticulum which release calcium. DAG remains in the plasma membrane and acts to activate protein kinase C.

soluble inositol polyphosphate headgroups and the 'headless' lipid diacylglyc-erol (DAG), which remains associated with the plasma membrane. Both products have biological activity. The inositol polyphosphates (principally the specific tri phosphorylated form, IP_3) mediated calcium release from the endoplasmic reticulum. From the perspective of mitogenic signalling, however, the principal player is the lipid product DAG.

Release of DAG by receptor-coupled phospholipase activity is a very common (but not ubiquitous) manifestation of signalling via diverse ligands with mitogenic activities. These include not only the small peptides that act via G-protein-coupled receptors but also receptors of the intrinsic tyrosine kinase class where phospholipase activity is activated by association of the enzyme with the receptor via an SH2 domain tyrosine phosphate-binding motif. Detailed examination of the composition of DAG released by mitogenic stimulation indicates that the process in reality involves several different phospholipases acting in a time-dependent manner which exhibit different substrate specificities.

Does DAG release have anything to do with cell proliferation? The major biochemical function of DAG, aside from its role in membrane biosynthesis, is to act as a co-factor for a family of plasma-membrane-associated protein kinases called the protein kinase Cs (PKC). These are protein kinases that exhibit calcium-dependent serine/threonine kinase activity for specific protein substrates. The PKCs are a large family of proteins which have a conserved core catalytic domain and a regulatory domain that binds calcium, DAG and plasma membrane lipids. Individual family members exhibit different patterns of tissue-specific expression, have different substrate specificities and different requirements for calcium and DAG composition for activation (Fig. 4.3). For certain members of the PKC family the calcium requirement is dramatically reduced in the presence of DAG which, under most conditions, leads to their activation.

Activation of PKC is sufficient to induce progress through the cell cycle and induction of DNA synthesis. The main evidence for this comes from a class of compounds called phorbol esters. Phorbol esters are essentially non-metabolisable pharmacological compounds that mimic the effects of DAG. Addition of PKC-activating phorbol esters to quiescent cells leads to the induction of DNA synthesis and cell multiplication. This seems to be primarily the result of PKC activation, since the mitogenic potency of phorbol esters correlates with their ability to activate PKC, and removal of PKC from cells blocks the ability of phorbol esters to induce DNA synthesis and cell multiplication. In addition, PKC mutants with constitutive catalytic activity have the ability to induce cell multiplication in transfected cells. It is, however, absolutely clear that PKC activation is not an obligatory requirement for progress through the cell cycle since suppression of PKC activation by, for example, mutation of phospholipase docking sites in intrinsic tyrosine kinase receptors, does not inhibit their ability to induce DNA synthesis.

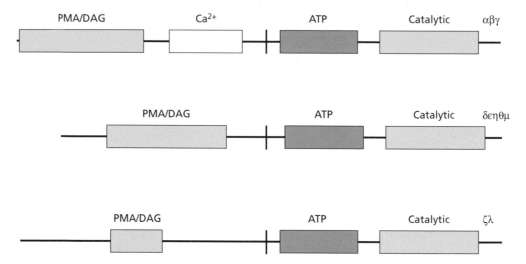

Fig. 4.3 Domain structures and subgroups of the protein kinase C family. Each member contains a conserved catalytic kinase domain at the C terminus and a lipid-binding domain (PMA/DAG) at the N terminus. The αβγ subfamily additionally exhibits a calcium-binding domain which renders enzyme activity sensitive to the intracellular calcium concentration.

Lipid kinases

One route to the identification of mitogenic signalling systems was to find proteins that bind to activated receptor complexes upon growth factor stimulation. This approach led to the discovery of a second class of enzymes that exploit inositol phospholipids to generate intracellular signals. These are the phosphatidylinositol-3′-OH kinases (PI3K). This is a family of enzymes which have the ability to phosphorylate membrane-associated inositol phospholipids in the 3′ position, thereby generating a range of new membrane lipid species (Fig. 4.4). PI3K activation is associated with mitogenic signalling as levels of 3′ phosphorylated lipid metabolites rise upon cell stimulation. This is due to the recruitment of PI3K to the plasma membrane upon receptor activation. This involves the docking of an SH2 domain containing an adaptor protein p85 to specific tyrosine residues in the phosphorylated receptor complex. The p85 subunit is bound to p110, which is the catalytic domain of PI3K. The net outcome of this is to bring the enzyme into the vicinity of its lipid substrates and activate the production of 3′ lipid metabolites.

There is a close association between receptor-mediated activation of PI3K activity and the induction of both progress through the cell cycle and processes involved in cell survival. For example, mutation of the p85 docking site in, for example, the PDGFR-beta receptor attenuates the ability of the receptor to induce cell cycle progression. Mitogenic signalling is also suppressed by introduction of 'dominant negative' versions of PI3K, which lack catalytic activity but

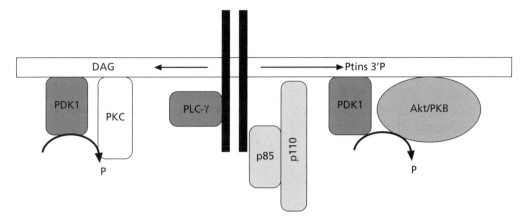

Fig. 4.4 Dual signal outputs from activated receptor tyrosine kinase (RTK) receptors. *Left*: Association of phospholipase C (PLC)-γ with activated receptors induces activation of PIP_2 hydrolysis and activation of protein kinase C (PKC) after phosphorylation by PDK1. *Right*: Association of the p85 subunit of PI3kinase with activated receptor recruits the catalytic p110 domain to the membrane, where it phosphorylates membrane inositol phospholipids in the 3′ position. The 3′ phosphorylated inositol phospholipids act as docking sites for proteins with PH domains. Recruitment of Akt to the membrane results in phosphorylation by PDK1 and concomitant activation of Akt activity.

retain the ability to associate with activated receptors. PI3Ks are also susceptible to inhibition by pharmacological compounds such as wortmannin. The application of wortmannin to stimulated cells can, in many cases, inhibit their ability to proliferate.

This cumulative evidence makes a persuasive case that the catalytic activity of PI3Ks is the key to understanding their mitogenic and cell survival functions. This focuses attention on the biological functions of 3′ phosphorylated phospholipds, since it may be presumed that they have some role in activation or recruitment of downstream mitogenic signalling pathways. As we might by now expect, 3′ inositol lipids are targets for membrane anchoring of proteins that contain specific inositol phospholipid binding sites with specificity for 3′ phosphorylated lipids. The pleckstrin homology (PH) domain is a structural motif found in a wide variety of proteins and is essentially a docking site for membrane inositol phospholipids (Fig. 4.5). Different PH domains have distinct phospholipid-binding specificities and PH domains have been located in a wide variety of proteins, including the cytoskeletal proteins pleckstrin (where the PH domain was first defined) and dynamin as well as more familiar proteins, such as some versions of phospholipase C. It is clear from this that the PH domain (and the functionally analogous FYVE domain) is a widespread mechanism for anchoring proteins to the cytoplasmic face of the membrane via modified phospholipids. This would lead to the suspicion that PI3K activity was potentially coupled into a wide range of cellular process that involve membrane–protein

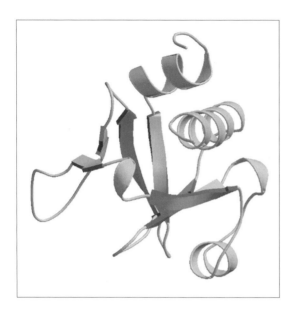

Fig. 4.5 Three-dimensional structure of the PH domain. Inositol phospholipid headgroups bind in the cleft formed between the core alpha helix and beta strands.

interactions. Indeed, the identification of an expanding family of enzymes with PI3K activity has shown that they are all involved in membrane transactions in, for example, controlling intracellular vesicular transport or endocytosis. In the context of mitogenic signalling via PI3K activation, we are therefore looking for some kind of protein with a PH domain (or equivalent) which is recruited to the membrane and activated in a PI3K-dependent manner.

The serine/threonine kinase AKT fulfils this prediction. The catalytic activity of AKT is elevated by activation of receptors with PI3K recruitment functions, and blocked by treatments such as wortmannin or receptor mutations which block PI3K activation. At least some of the biological effects of PI3K activity on cell proliferation and cell survival can be emulated by expression of activated mutant forms of AKT showing that, of the many downstream consequences of PI3K function, the mitogenic and survival signalling pathways are predominantly channelled through AKT activation. Mutation of the amino terminal PH domain of AKT impairs its activation by PI3K, indicating that recruitment to the membrane via 3' inositol phospholipids is a crucial feature of AKT function. Membrane recruitment *per se* is, however, insufficient to fully activate kinase activity. This also requires phosphorylation of the AKT protein in the activation loop of the kinase core which is mediated by another PH domain containing kinase PDK1 (Fig. 4.4). Thus, the activation of PI3K results in AKT activation directly via membrane anchoring and indirectly via an activating kinase.

We therefore see that several different signalling pathways that emanate from active signalling receptor complexes output into the activation of specific intracellular protein kinases, including specific forms of PKC and AKT. This is achieved by mechanisms that rely on the receptor-coupled enzymatic generation of specific membrane phospholipids and which leads directly or indirectly to membrane docking of the 'output' kinase and concomitant phosphorylation of protein substrates. As with the process of docking to activated receptors, the use of protein motifs with lipid-binding activities is a key element in the design of the pathway.

The Ras/Raf/ERK pathway

The molecular elegance of the G protein as biochemical switching mechanism in mitogenic signalling is exemplified by the functions of the Ras family of G proteins. The *Ras* genes were originally discovered by virtue of their ability, when subjected to mutations which inactivate their GTPase activity, to activate mitogenic signalling pathways and cell proliferation. The origins of these mutations are discussed more fully in Chapter 7. However, from our knowledge of how G proteins work, we must assume that Ras proteins in this situation are permanently engaged with some type of 'downstream' signalling process. By the same token, we must also assume that the normal catalytically active forms of Ras are coupled, by some means, into mitogenic receptor activation.

Regulation of Ras activity

Ras proteins are membrane associated via an acylated lipid anchor in their C-terminal domain. The membrane anchoring of Ras proteins is essential for their mitogenic signalling activity since mutation of the membrane anchor leads to disruption of Ras signalling. It seems from this that the localisation of Ras proteins in the membrane is an important element in their mitogenic functions.

A second essential feature of the Ras family is that their GTPase activity and GDP/GTP exchange functions are controlled by the association of Ras with other cellular proteins. These are the GEFs (guanine nucleotide exchange factors), which switch *Ras* on, and the GAPs (GTPase activating proteins), which effectively switch *Ras* off (Fig. 4.1). Based upon our knowledge of receptor activation mechanisms, we might come to the conclusion that the upstream activation of Ras effector functions are connected to regulation of the activity of Ras GEFs and GAPs.

The identification of Ras family effectors has been greatly aided by the use of genetic screens in organisms such as *Drosophila* and *C. elegans* for mutations which disrupt or modify signalling via the Ras pathway. One such gene is *Sos* (for son-of-sevenless). *Sos* codes for a GEF that catalyses the exchange of GDP for GTP on Ras leading to Ras activation. Genetic disruption of *Sos* in flies leads to inhibition of Ras-dependent signalling pathways, suggesting that *Sos* is a significant mediator of Ras-dependent signalling pathways. *Sos* is activated by association of the protein with an activated receptor complex through an adaptor protein called Grb-2 (for growth factor receptor bound). Grb-2 is a simple molecule which contains two SH2 domains, thereby binding to phosphorylated

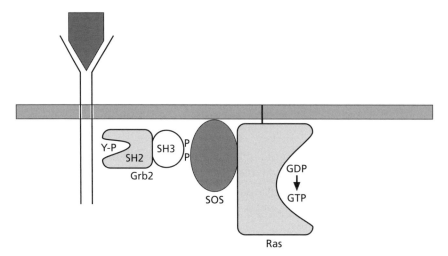

Fig. 4.6 Activation of Ras by activated receptor tyrosine kinase (RTK) receptors. The SH2 domain containing protein Grb2 binds to an activated receptor complex. Grb2 recruits the guanine nucleotide exchange factor SOS which associates with Ras and switches Ras to the active GTP-bound form.

tyrosines in the receptor complex, and an additional protein motif called the SH3 domain, a motif which interacts with polyproline sequences in target proteins. SOS binds Grb-2 via the SH3 domain interaction and can be found complexed with this adaptor in the cytoplasm of resting cells. Upon cellular stimulation of receptor tyrosine kinases, the Grb-2–SOS complex is recruited to the tyrosine phosphorylated receptor via the SH2 domains of Grb-2, allowing interaction between SOS and Ras, which is membrane localised via an acylated C-terminal domain (Fig. 4.6). The Grb-2 adaptor is essential for SOS activation, since mutation of the fly homologue of Grb-2 completely abrogates signalling through the Ras system. Depending upon the receptor involved, the Grb-2–SOS complex can be also recruited to the membrane via the use of another SH2 domain adaptor protein (SHC) or via the membrane-bound intermediate FRS2 (FGF receptor substrate 2), which is phosphorylated by an activated receptor kinase. However, whatever the route chosen, it is the membrane recruitment of SOS via the Grb-2 adaptor that is the key element in the activation of Ras, since artificial targeting of SOS to the membrane via addition of a lipid anchor activates Ras function.

The activation of Ras effector function by GEFs is reversed by the activity of Ras-GAPs, which attenuate Ras signalling via activation of the intrinsic Ras GTPase activity. A number of Ras GAP proteins have been identified by a variety of means; these include p120-GAP, GAPm and neurofimbrin. The GAPs also associate with receptor complexes via SH2 docking sites. This leads to the idea that the ultimate state of Ras activity in the cell is dependent upon the relative activity of GAPs and GEFs and upon the amount of Ras protein present in cells.

Downstream effectors

The G protein switch model predicts that the biological effects of activation of Ras by GTP loading is mediated by the association of active Ras with downstream effectors. Since Ras is tightly bound to the internal face of the membrane via its acyl tail, it might also be predicted that association of a putative effector would lead to the formation of a membrane-bound signalling complex. The most important downstream target for Ras effector activities is the serine/threonine kinase Raf. Activation of the Raf kinase is required for the mitogenic activities of Ras proteins. Construction of activated mutant forms of Raf bypasses the need for Ras activation, and inhibition of Raf kinase activity by, for example, the construction of dominant negative forms of Raf can abrogate mitogenic signalling via Raf activation.

Raf interacts directly with the Ras/GTP complex but does not bind the Ras/GDP complex. Thus, activation of Ras to the GTP-bound state recruits Raf to the plasma membrane. Mutation of the amino acid residues in Ras that interact with Raf block the ability of Ras to elicit a mitogenic signal. Membrane association seems, by some means, to be the major mechanism for activation of Raf kinase activity, since Raf can be activated by artificial recruitment to the plasma membrane by addition of a lipid anchor. However, activation of Raf kinase activity also requires phosphorylation of key residues in Raf and this may be mediated by other proteins that associate with the Ras–Raf complex. Amongst the other components of this complex is a protein 'scaffold' protein 14–3-3 (named after its relative mobility in gels). The 14–3-3 protein binds to Raf and is required for Raf kinase activity in a number of different systems. Loss of 14–3-3 function by mutation in yeast blocks the ability of Ras to activate Raf kinase activity. The 14–3-3 proteins appear to enhance Raf kinase activity by stabilising phosphorylated forms of the protein, perhaps by prohibiting access to Raf phosphatases.

What is the cellular target for the active form of the Raf kinase? A number of targets of Raf have been identified, including GEFs for other G proteins and, confusingly, PI3K. The overwhelmingly important target from the cell cycle perspective is, however, a highly conserved cascade of three protein kinases called the MAPK (mitogen-activated kinase) cascade (Fig. 4.7). This cascade comprises Raf (or MAP kinase kinase kinase; MKKK) itself at the top of the pathway, which phosphorylates the next member of the pathway, MEK (or MAP kinase kinase; MKK). MEK is a dual specificity kinase which can be phosphorylated on both tyrosine and threonine residues in a sequence-specific manner. The 'end' of the cascade is the MEK substrate, which is a serine/threonine-specific kinase, ERK (or MAP kinase; MK). It is the activation of ERK which is required for the ability of both Ras and Raf to induce DNA synthesis and cell proliferation in resting cells.

The MAP kinase cascade is a highly conserved biochemical signalling unit that is employed, in various guises, in a wide variety of signalling systems. The basic three-kinase unit, involving a suite of structurally related enzymes, is used

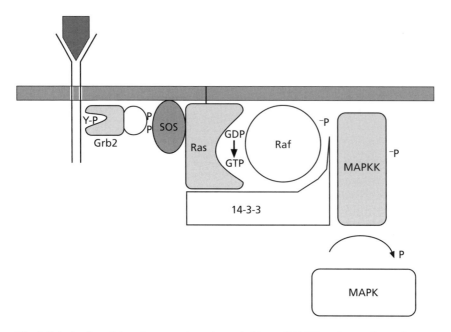

Fig. 4.7 Activation of the mitogen-activated protein kinase (MAPK) pathway. The exchange of GDP for GTP on Ras leads to the formation of a complex with downstream effectors. These include the scaffolding protein 14-3-3 and the first member of the MAPK pathway, Raf. Activation of Raf leads to activation of MAPKK, which in turn phosphorylates MAPK. MAPK can dissociate from the complex and translocate to the nucleus.

for signalling responses to physiological stimuli as diverse as the mating response in yeast and the stress response to osmotic shock or protein synthesis inhibition in mammalian cells.

A key theme in the activity of the MAPK module is the cellular localisation
of the component elements. The first enzyme, MKKK, is invariably membrane
bound; the second component, MKK, appears to shuttle between membrane
and cytoplasmic compartments; and the third enzyme, MK, has the ability,
under certain conditions, to enter the nucleus and phosphorylate target substrates. As we shall see in Chapter 5, this is the essential feature of the involvement of the mitogenic MAPK cascade (i.e. Raf, MEK and ERK) in the induction
of progression through the cell cycle. The idea of relocalisation of components
of the MAPK module is reinforced by genetic studies of MAPK modules in yeast.
These reveal the existence of a specific class of scaffolding proteins which are required for regulation of the MAPK pathway through control of the physical association of the components. Scaffolding proteins, such as the mammalian
protein MP1, bind to component enzymes of the MAPK pathway and enhance
their activity. The mechanism by which this is achieved is not wholly understood. One possibility is that these proteins act to increase the catalytic efficiency of the pathway by bringing the enzymes into close physical proximity,

and/or physically shielding them from cytoplasmic phosphatases. It is certain, however, that the activity of these scaffolding proteins must be subject to dynamic regulation, perhaps by other protein kinases which permit the component elements of the MAPK pathway to relocate within the cell.

Intracellular signal integration

Our account of receptor-mediated signal transduction has focused on the mechanism by which three specific types of protein kinase systems are activated by receptor activation. These are the PKC pathway, the PI3K/AKT pathway and the MAPK pathway. On the surface it would appear as though these are all essentially independent outputs and that an individual receptor, by virtue of the docking site 'code', can simply dial into one or more pathways by having the appropriate docking sites. The action of a particular receptor is therefore dictated by the combination of downstream kinases that are activated.

There are, however, reasons to suppose that things are significantly more complicated. This in part arises from the existence of intersection points between these different signalling pathways. Thus, for example, the MAPK module is a target for regulation by a variety of pathways other than the Ras–Raf system. Of considerable significance is the finding that a major substrate for the DAG-activated forms of PKC is the MAPK pathway. Different forms of PKC have been found to phosphorylate both Raf and MEK in phorbol ester-stimulated cells, leading to a potentiation of their activity. Indeed, it is likely that regulation of the MAPK cascade is the predominant mechanism by which cell cycle progression is induced by PKC activity. There are also strong connections between the PI3K and PKC pathways since the enzyme PDK1, which is required to potentiate the activity of AKT, also increases the catalytic activity of PKC by phosphorylation of the activation loop. This indicates that the activity of one signalling pathway is modulated by the concurrent activity of another pathway.

A second issue is the kinetics of signalling pathways. This is most obviously manifest in control of the cellular localisation of the MAPK pathway. It has been frequently noted that the duration of activation of signalling pathways is influenced by the receptor at the top of the pathway. Thus, activation of the MAPK pathway in cultured cells by agents such as EGF results in a relatively brief elevation of ERK activity and a failure of the enzyme to undergo nuclear translocation. Stimulation of the ERK pathways by activation of other intrinsic tyrosine kinase receptors, such as the nerve growth factor (NGF) receptor in PC12 cells, leads to prolonged ERK activity and this is associated with translocation of active ERK to the nucleus. Since dynamic relocalisation of the MAPK pathway is a critical aspect of its activity, the influence of modulating the kinetics of signalling may have quite a profound consequence for induction of cell cycle progression.

The final complication arises from considering the simultaneous activation of multiple signalling pathways. If different signalling pathways can intersect and modulate each other's behaviour, it is easy to envisage a situation where one pathway can function only when another is also active. This would lead to a requirement for the simultaneous action of two ligands to elicit a mitogenic response. This is most obviously seen as the requirement by particular cells (such as 3T3) to be attached to a substrate in order to respond to mitogenic signals. Indeed, it is uncoupling the requirement for attachment from mitogenic receptor signalling that is one of the manifestations of TGFB action described in Chapter 2. This situation indicates that the process of attachment must involve the activation of signalling pathways that need to be active for mitogenic signalling systems to operate. Cell attachment to a substrate is mediated by the interaction with a specific family of heterodimeric cell surface receptors, the integrins, with protein components of the extracellular matrix (ECM). Prominent examples of integrin ligands are proteins which promote cell adhesion, such as fibronectin and laminin. Integrin activation by ECM ligand engagement is known to activate a number of intracellular pathways that impinge upon the organisation of the cytoskeleton. These cytoskeletal assembly pathways are controlled by a family of Ras-related G proteins including Rho and Rac. Activation of Rho and Rac by unknown means related to integrin engagement leads to dynamic reorganisation of cytosketal components. This reorganisation process is in some way coupled to the ability of mitogenic signals to induce cell cycle progression. The mechanistic basis of this intersection between adhesion-dependent and mitogenic signalling pathways is almost totally obscure and further confounded by the fact that in most experimental studies the signalling pathways activated by both systems appear to exhibit strong similarities. However, the principles that have guided our analysis of mitogenic signalling in this chapter most surely indicate that the inter-relation between these pathways will revolve, in some way, around the dynamic formation of protein complexes within specific cellular compartments.

Intracellular signalling in perspective

Receptor activation leads to stimulation of multiple types of intracellular signals. Despite the intrinsic (and necessary) complexity of mitogenic signalling pathways, two strong principles emerge. The dynamic assembly of specific protein complexes in specific cellular locations is the central molecular mechanism of intracellular signalling. All signalling systems with a mitogenic involvement output into the activation of various types of protein kinases—it is the function of these kinases that accordingly lies at the heart of the next phases of cell cycle progression.

Chapter 5: Gene Expression

Introduction

So far we have seen how growth factors, by interacting with cell surface receptors, initiate a cascade of biochemical events which are characterised by relocation of proteins within the cell and activation of enzyme activity. Where does this flurry of activity lead? The essential answer is that it culminates in the activation of gene expression and is the appearance of new gene products, as a result of the metabolic processes of signalling, that play an essential role in the progression into S phase and completion of the cycle.

It is actually by no means obvious that cell cycle control should involve new gene expression. It might be possible to envisage schemes by which fluctuations in the activity of pre-existing enzymes could lead to DNA synthesis and cell division. However, events such as DNA synthesis and mitosis require complicated biochemical machinery to be activated in a coordinated manner. The use of gene regulation as a mechanism for precisely organised activation of many different proteins is obviously attractive. Furthermore, the induction of specific sets of genes also provides a potential explanation for the observation that cell cycle progress switches from processes dependent upon external signals to processes which are purely cell autonomous at the restriction point. Following this line of argument requires that we are able to answer a number of questions: is gene expression activated by growth factor signalling? Is new gene expression required for progress through the cell cycle? If the answer to both questions is affirmative (which of course it is), then we need to explain how receptor activation at the membrane leads to transcriptional activation in the nucleus. That is the purpose of this chapter.

Induction of gene expression by growth factors

A superficially simple experiment immediately reveals that new gene expression results from growth factor signalling. Quiescent cells are exposed to growth factors and, at various time points afterwards, ribonucleic acid (RNA) is extracted from the cells and examined for the presence of transcripts encoded by a panel of genes. The results of such an experiment are shown in Fig. 5.1. This clearly shows that the expression of certain genes changes in the period after growth factor stimulation: in particular, some transcripts are absent from quiescent cells and appear after stimulation.

A closer look reveals that individual genes exhibit characteristic kinetics. Rather then every gene being switched on at once, there appears to be distinct

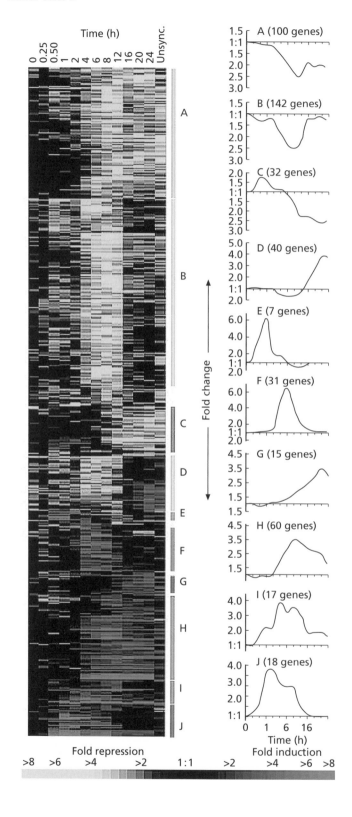

ordering of events. Some genes, such as *c-fos*, are activated within a few minutes of growth factor stimulation: genes of this type represent the early class of genes. Other genes, such as collagenase, are switched on within a few hours: these are called the intermediate class of genes. Finally, some genes appear after 12–16 hours, at around the time DNA synthesis is initiated. These are the so-called late genes. Amongst the family of late genes are many involved, directly or indirectly, in the process of DNA replication itself. These include enzymes involved in nucleotide metabolism, such as dihydrofolate reductase, and components of the DNA replication machinery, such as the γ subunit of DNA polymerase. Another characteristic of the late class of genes is that they appear after the restriction point has passed. This implies that these genes must be activated indirectly by growth factor signalling, since a requirement for receptor activation has passed.

Not only does the appearance of new transcripts in response to stimulation exhibit specific kinetics but also the expression of some genes appears very transient (Fig. 5.1). Transcripts encoding the early gene *c-fos*, for example, disappear within a couple of hours of their appearance and are not present in the cell at the same time as late genes. This implies that there must exist at least two mechanisms to control these genes: one to switch them on and one to switch them off again. Indeed, by definition, every gene whose expression is regulated by growth factor signalling must be dependent on some time-dependent processes, since they are all switched off as cells enter quiescence.

The transient kinetics of growth factor-induced gene expression and the overt ordering of gene expression events suggests that the expression of some genes might be dependent upon the prior expression of others. In other words, amongst the early class of genes might exist some whose function is to activate the intermediate class of genes, and amongst the intermediate class of genes are some whose function is to induce the late class of genes. Following this point further it might be thought that, since late genes are activated after the restriction point, they might be regulated by intermediate genes which are induced before or around the time of the restriction point. This leads to the idea that the restriction point represents a point where a critical set of genes involved in the expression of components of the DNA replication machinery are induced.

The idea that the different temporal classes of growth factor-induced genes are interdependent can be simply tested. If the expression of certain genes is de-

Fig. 5.1 Alterations in gene expression in response to serum stimulation. Quiescent human fibroblasts were exposed to serum and RNA samples taken at various time points thereafter. Fluorescently labelled cDNA derived from the RNA samples was then hybridised to a panel of genes immobilised on a glass slide (from top to bottom). The amount of mRNA for each arrayed gene is reflected in the intensity of the fluorescent signal. It will be seen that the panel of selected genes show characteristic alterations in expression at varying time points after serum stimulation. (From Iver *et al.* (1999) *Science* **283**, 83–87.)

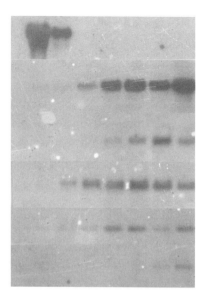

Fig. 5.2 Effect of protein synthesis inhibition on gene expression. Quiescent cells were stimulated with serum or treated with the protein synthesis inhibitor cyclohexamide. RNA samples were prepared at various time points, separated by gel electrophoresis and probed with labelled cDNAs corresponding to genes that are induced by growth factors with different kinetics. It will be seen that inhibition of protein synthesis leads to the appearance of growth factor-inducible transcripts in the absence of stimulation.

pendent upon the prior expression of others, then their induction should be blocked by the application of protein synthesis inhibitors after growth factor stimulation. At first sight the results of such experiments (Fig. 5.2) produce a counter-intuitive result. Early gene activation is unaffected by inhibition of protein synthesis and, in fact, levels of transcripts actually increase as a result of blocking protein synthesis. This unexpected finding is in part due to the suppression of proteins that are required for the destruction of transiently activated transcripts such as c-fos. Nevertheless, the fact that early gene expression is unaffected by blocking protein synthesis indicates that this category of genes must be activated by the release of some kind of latent mechanism; i.e. all the components required for the expression of *c-fos* are already present in the cell during quiescence but in an inactive state. One function of receptor activation would therefore be to activate the elements required for expression of the early class of genes. In contrast to early genes, the intermediate and late classes of genes are suppressed by inhibition of protein synthesis after growth factor stimulation. This suggests that their expression must require the production of functional protein from transcripts of the early class, amongst whose members must be factors which act to induce intermediate and late gene activation. This suggests that there exists a cascade of gene expression which is set in motion by growth factor signalling acting to release some latent mechanism of early gene activation. Amongst early genes are transcriptional activators of intermediate genes and amongst the intermediate genes are transcriptional activators of the late genes (Fig. 5.3). It will be seen from this model that direct dependancy on concurrent receptor signalling is lost as we move from the intermediate to late class of genes.

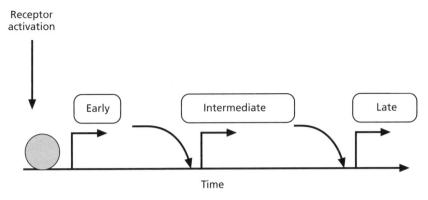

Fig. 5.3 A simplified model of the sequence of gene activation in response to growth factor stimulation. Induction of early genes, which encode transcription factors in response to growth factor stimulation, leads to the expression of genes with intermediate characteristics. Amongst the intermediate genes are transcription factors which activate the expression of genes with late temporal characteristics.

The requirement for gene expression in cell cycle progression

The data discussed above clearly show that gene expression is a byproduct of growth factor signalling but do not directly establish that this cascade of gene expression is required for induction of DNA synthesis and progress through the cell cycle. This can be tested in many ways but perhaps the most elegant experiment is one which showed that exposure to growth factors produces an RNA product whose function is to trigger the induction of genes required for the induction of DNA synthesis (Fig. 5.4).

Quiescent cells were exposed to PDGF for sufficient time to induce passage through the restriction point but some hours prior to the onset of DNA synthesis. The PDGF was washed out, the cells were enucleated and fused to quiescent cells that had never been exposed to PDGF. These hybrid cells, whose nuclei had never been directly exposed to PDGF-mediated signalling, underwent DNA synthesis on schedule. The ability of PDGF-treated cell cytoplasm to induce DNA synthesis in a virgin nucleus was completely blocked by exposure of the cells to inhibitors of gene transcription during the period of exposure to PDGF. However, exposure of cells to inhibitors of protein synthesis during exposure to PDGF did not inhibit the induction of DNA synthesis in the fused cell nuclei. The clear conclusion to be drawn from this experiment is that PDGF-mediated signalling induces the appearance of an RNA species whose function is to induce the sequence of events leading to DNA synthesis, including the induction of genes required for DNA replication. Once this RNA species appears there is no further requirement for PDGF signalling for further progress through the cell cycle. This concept sounds like a redefinition of the restriction point, which

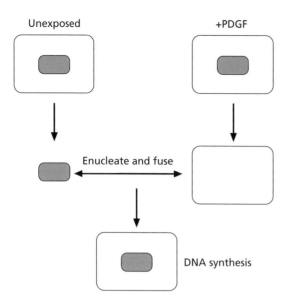

Unexposed

+PDGF

Enucleate and fuse

DNA synthesis

Fig. 5.4 Induction of gene expression is required for the induction of DNA synthesis. Quiescent cells were either exposed to platelet-derived growth factor (PDGF) or left unexposed. After four hours, both sets of cells were enucleated and the nuclei from unexposed cells fused with the cytoplasm of cells exposed to PDGF. The hybrid cells were exposed to radiolabelled nucleotides and the induction of DNA synthesis measured. The nuclei from the unexposed cells underwent DNA synthesis on the cytoplasm of the cells exposed to PDGF. This did not occur if gene transcription was blocked during the period of exposure to PDGF. This indicates that gene transcription induced by exposure to PDGF is required for the onset of DNA synthesis.

again leads to the idea that the restriction point may, in fact, represent the time at which the expression of these critical genes is induced.

The identity of growth factor-inducible genes

It follows from the conclusion that growth factor-induced genes play a key role in inducing cell cycle progression that the identity and function of genes induced by growth signalling might be of interest.

There are broadly two strategies for identifying genes induced by growth factor signalling. The first is to test known genes and the second is to clone genes purely on the basis of their growth factor inducibility. Both strategies converge on a common set of conclusions.

Only a limited number of genes are regulated, directly or indirectly, by growth factor signalling. Current estimates place the number in the order of a few hundred (or about 0.1% of the total number of genes in mammals such as humans). This is obviously a relief. It eases the identification of genes with key functions and also proves that induction of gene expression by growth factors is most unlikely to represent a generic response to signalling, but almost certainly has some functional significance.

Amongst the genes that have been identified are many which appear to have little obvious connection to the business of cell cycle progression. In particular, many growth factor-inducible genes of the intermediate type encode extracelluar proteases such as collagenase or protease inhibitors, for example, tissue inhibitor of metalloproteinases (TIMP). The reason for this almost certainly reflects the fact that the majority of such studies on gene regulation have been performed with fibroblast cell lines such as 3T3, which, as described in Chapter 1, can be regarded as undergoing a form of 'wounding' response when exposed to growth factors *in vitro*. However, it turns out that these genes have been extremely useful in unravelling the molecular details of gene regulation, as will be described below.

It is clear from what has been stated already that a useful place to start filling in the molecular details of growth factor-inducible gene expression would be the early genes, since they must, by virtue of the timing of expression and latent control mechanisms, be most closely linked to the receptor-mediated signals described in Chapters 3 and 4. The standard example of an early gene is *c-fos*. The discovery of *c-fos* will be described in more detail in Chapter 7. For the moment, it simply needs to be known that *c-fos* is expressed as a nuclear protein of about 40 kDa and is induced within minutes of growth factor stimulation, expression peaks after about one hour and then rapidly declines so that, under most circumstances, expression has returned to basal level within three hours. We already know that the activation of *c-fos* gene expression must involve activation of some latent process since its expression is unaffected by inhibition of protein synthesis.

How is the transcription of the *c-fos* gene activated? This question can be approached by dissection of transcriptional regulatory elements in the *c-fos* promoter; these confer growth factor inducibility upon the gene. This is achieved by introducing the *fos* gene, along with 5' flanking sequences, into cells and then testing for regions which are required for growth factor inducibility by systematic deletion of 5' flanking sequences. This type of analysis has defined several different elements in the 5' region of the *c-fos* gene which are required for growth factor inducibility and which can also confer growth factor inducibility upon a heterologous gene.

An especially effective regulatory element has been defined in the *c-fos* promoter region which confers upon foreign genes the property of inducibility by serum. This is the serum response element or SRE. How does the SRE DNA sequence elicit serum inducibility? The most likely mechanism would be that the DNA sequence interacts with a nuclear transcription factor which activates gene expression. Indeed, a transcription factor called the serum response factor (SRF) was isolated; this bound the SRE DNA sequence with high affinity. SRF is a member of the larger MADS box family of transcription factors. Perhaps surprisingly, the SRF is present in quiescent cells and its level of expression changes little during the cell cycle. Furthermore, it can be shown that the SRE site is bound to the SRF in quiescent cells and remains bound after stimulation. It seems therefore that induction of *c-fos* must require some modification of the activity of SRF for gene induction to occur. This occurs by the association of the SRF with another nuclear protein called Elk-1 (or ternary complex factor; TCF) after serum stimulation. It is the formation of the Elk-1/SRF protein complex on the SRE DNA sequence which provides the trigger for transcription of the *c-fos* gene (Fig. 5.5). The problem therefore reduces to the question: What controls

Quiescent

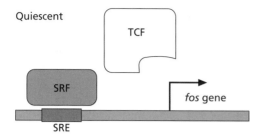

Growth factor stimulation

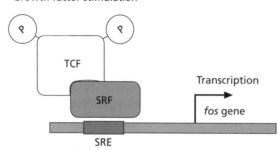

Fig. 5.5 Activation of *fos* gene expression in response to receptor signalling. In non-stimulated cells the serum response element (SRE) DNA element is bound by the serum response factor (SRF). The ternary complex factor (TCF) is phosphorylated by mitogen-activated protein kinase translocating to the nucleus upon receptor activation. Phosphorylated TCF forms a complex with the SRF which induces activation of *fos* gene transcription.

the association of Elk-1 with the SRF and what is the point of intersection between Elk-1 and receptor-mediated signalling pathways?

Elk-1 protein associated with the SRF is phosphorylated on serine residues and Elk-1 phosphorylation can be stimulated by a variety of different approaches, which have in common the ability to activate the MAP kinase pathway described in Chapter 4. In fact, it is possible to show *in vitro* that Elk-1 phosphorylation by MAP kinase occurs on exactly the same residue that is phosphorylated in serum-stimulated whole cells. The conclusion seems clear; Elk-1 is phosphorylated by MAP kinase, which is activated by the Ras/Raf pathway. Phosphorylation of Elk-1 permits it to complex with SRF and the formation of the Elk-1/SRF complex triggers *c-fos* gene transcription (Fig. 5.5).

This is a satisfying account of how receptor-mediated kinase cascades can feed directly into the activation of gene expression and is also a significant oversimplification of the true state of affairs. First, it takes no account of the chromosomal environment of the *c-fos* gene in quiescent cells. It now seems clear that some changes to histone packing, probably by acetylation or phosphorylation of histones, must also occur for gene induction to take place. These may be mediated by other receptor-mediated pathways. Second, the activity of the SRF can be modified by a number of other signalling pathways, especially the poorly understood systems which lie downstream of the Rho family of small GTPases. Third, the *c-fos* promoter contains a number of other DNA elements, aside from the SRE, which control growth factor inducibility in both positive and negative ways. The 'real' control of the *c-fos* gene in whole cells therefore almost certainly involves the integrated modification of all the regulatory elements and the chromosomal environment by the convergence of a number of different signalling pathways.

Genes downstream of *c-fos*

Whilst *c-fos* activation provides a good example of how signalling pathways feed into transcriptional activation, it also turns out that *c-fos* provides very useful information on the regulated control of the intermediate category of genes.

The same approach that led to delineation of growth factor-regulated elements in the *c-fos* promoter, namely deletion of DNA elements in the vicinity of the 5′ end of the gene, can be applied to understanding the control of intermediate gene expression. As described above, many intermediate genes, such as collagenase, may have little direct relevance to cell cycle control but have proved to be valuable test cases for identifying potential transcriptional regulatory events that lie upstream of intermediate gene induction.

Deletion analysis of the collagenase promoter reveals a minimal DNA sequence for induction of collagenase by phorbol esters and growth factors. This DNA element was termed the TRE (for TPA regulatory element). Multiple copies of the basic TRE sequence could, when ligated to a heterologous promoter, confer growth factor inducibility on a heterologous gene. The sequence of the TRE was very similar to the known recognition sequence for a yeast transcription factor called GCN4. This suggested that there may be some similarity between the protein(s) that interact with the TRE and the GCN4 protein from yeast. Studies of proteins bound to the TRE in quiescent and stimulated cells showed that the TRE was unoccupied in quiescent cells but became bound by proteins within an hour of stimulation. In addition, the induction of collagenase could be prevented if cells were treated with protein synthesis inhibitors during the period of stimulation. This suggests that activation of the collagenase gene required *de novo* synthesis of a protein rather than direct activation of a latent activity by the signalling machinery. These observations led directly to the idea that activation of collagenase is due to occupation of the TRE by proteins that were rapidly induced by growth factor signalling, namely a member of the early class of growth factor-inducible genes.

Given that one of the best known early genes was *c-fos*, it was an obvious question to ask whether *c-fos* could be one of the proteins that associate with the TRE following growth factor stimulation. The answer to the question was that it was. The c-fos protein binds to the TRE in stimulated cells and it does so in partnership with another protein called c-jun, a DNA-binding nuclear protein of about 39 kDa which, like c-fos, is subjected to post-translational phosphorylation. As was previously speculated, the DNA-binding domain of c-jun proved to show significant sequence homology to the yeast transcriptional activator GCN4. The idea emerges, therefore, that the collagenase gene is transcriptionally silent until the TRE sequence is bound to a transcriptional activation complex. This complex is formed by hetero-dimerisation of c-fos and c-jun and its formation is therefore dependent upon prior transcriptional activation of the *c-fos* gene.

How do c-fos and c-jun dimerise? The mechanism involves a very elegant structural motif called the leucine zipper. This is an alpha helical module, found not only in c-jun, c fos and GCN4 but also in a wide variety of other transcription factors. The sequence of the helical domain is such that the exposed leucine residues are brought into register down one face of the helix (Fig. 5.6). This enables two zipper domains to dimerise by the opposing leucines interdigitating in the manner of teeth in a zip. In the case of the fos/jun heterodimer, the creation of the zipper dimer brings the adjacent DNA-binding domains into register to create a full high-affinity sequence-specific binding site. Zipper-generated dimers exhibit exquisite specificity in pairing due to the arrangement of residues flanking the leucines. It can be experimentally shown that dimerisation is essential for the fos/jun complex to elicit transcriptional activation by

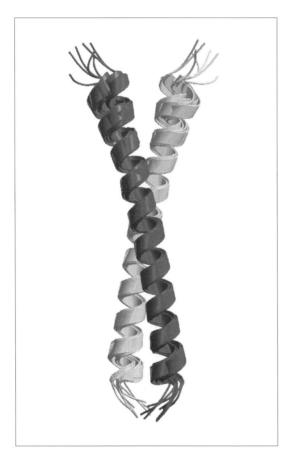

Fig. 5.6 The three-dimensional structure of the leucine zipper. Fos and Jun form a heterodimer via association of their complementary zipper domains. These are composed of long alpha helices which contain leucine residues along one face of the helix.

mutating the leucine residues in the partner zipper domains. This can be further extended by showing that the essential function of the zipper domain is to induce dimerisation: this is achieved by swapping the zipper domains from other proteins such as GCN4. This modular format of zipper domain transcription factors represents a further example of how significant diversity in recognition can be brought about by mixing and matching closely related proteins.

This account of intermediate gene activation by early gene products conceals a significant complexity. Both *c-fos* and *c-jun* are actually members of a larger family of related genes which are able to form combinatorial pairings via the zipper mechanism. This has two implications. One is that, as a result of differences in DNA sequence recognition preferences, the basic fos/jun dimer format can be employed to transcriptionally activate a variety of different genes with different recognition sites in their promoters. This means that the fos/jun binding site in the collagenase promoter is a prototype example of intermediate gene regulation rather than a signature sequence characteristic of all intermediate growth factor-inducible genes. In addition, some members of the extended jun family have defective DNA-binding domains. This means that when a dimer is formed with a partner, DNA binding is blocked. These jun variants are in effect dominant negative inhibitors of gene transcription. The implication of this is that, in principle, the fos/jun transcriptional regulatory mechanisms can be used in a wide variety of circumstances to both activate and inhibit gene expression.

The collagenase promoter which, compared with other intermediate genes, is quite simple in design, contains many more regulatory elements than the TRE. Some of these act to amplify or diminish TRE-dependent transcriptional activation. As with the *c-fos* promoter discussed earlier, the actual expression of the collagenase gene is a consequence of the integrated activity of all these elements acting in concert. Moreover, the account of collagenase gene regulation via the fos/jun heterodimer is based upon evidence derived in large measure from studies of gene regulation in fibroblast cell lines. It is more than likely that the complexity of the transcriptional regulatory elements found in intermediate genes reflects additional control elements required to control gene expression in a cell-type specific manner.

Despite these caveats, a reasonable working hypothesis which describes the relationship between receptor-mediated signalling to gene expression can be defined.

Receptor-mediated signalling feeds into activation or inhibition of latent gene-control proteins which induce the expression of genes of the early class. Frequently, phosphorylation and relocation of specific control proteins play a key role in this process. Further examples of this type of mechanism have already been described in Chapter 3 in the form of the STAT and SMAD transcription factors which are associated with inactive receptor and become released upon activation of receptor-associated kinase activity, translocate to the nucleus and activate downstream genes. Another type of mechanism is illustrated

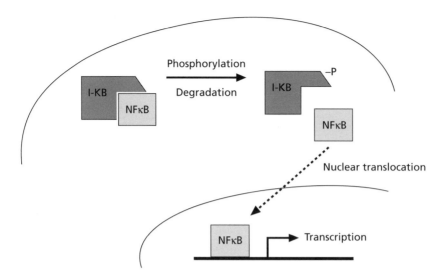

Fig. 5.7 Induction of gene expression by NFκB. In unstimulated cells, NFκB is bound to I-KB which occludes the nuclear localisation sequence of NFκB. The activation of receptor signalling results in phosphorylation of I-KB; this dissociates from NFκB which translocates to the nucleus and activates gene expression.

by the transcription factor NFκB, which exists in the cytoplasm as an inactive complex with an inhibitor IKB. Activation of cytoplasmic kinases by receptor-mediated signalling leads to breakdown of the complex and relocation of the transcription factor to the nucleus (Fig. 5.7).

Amongst the early category of genes are transcription factors whose expression leads to the induction of intermediate genes, amongst whose members are presumably other transcription factors that act on further downstream genetic targets. The key point is that after activation of the early category of genes, there is no continued requirement for receptor-mediated signalling to occur. This provides a mechanistic explanation for the experimental result on hybrid cells described in Fig. 5.4. What this account does not reveal is the identity of the gene(s) whose action commits the cell to execution of DNA synthesis. A partial answer to this question will emerge when the molecular mechanisms controlling entry into DNA synthesis are defined in the next few chapters.

Chapter 6: The Cell Cycle Engine

Introduction

Previous chapters have led us to the restriction point: that experimentally defined event in the cell cycle where cells commit to completion of the cycle without further need for an input from receptor-mediated signalling. In this chapter we shall be concerned with the biochemical mechanisms that dominate the remainder of the cell cycle through to mitosis. This sequence of events is often called the 'cell cycle engine', a core mechanism that executes the visible landmark processes of the cell cycle S phase and mitosis. The cell cycle engine has another important feature: the multiplication of every species of eukaryotic cell from yeast through to humans involves the same set of processes. It will be of little surprise to learn therefore that the biochemical machinery that comprises the cell cycle engine exhibits extreme conservation. This, as we shall see, provides a powerful experimental lever for prising open the mechanism.

The molecular mechanisms that comprise the cell cycle engine exhibit a number of key properties.

1 The engine involves an ordering of different events in time. Thus, S phase precedes mitosis in the normal cell cycle rather than the two processes happening at random or concurrently. If one purpose of cell division is to deliver an intact genome to the next generation, this feature is clearly essential. The exception to this rule, intriguingly, is meiosis where two rounds of chromosome segregation and cell division occur without an intervening round of DNA synthesis.

2 The engine involves control points in which certain events are prohibited from happening before others have been completed. Thus, S phase must be completed before mitosis is initiated, otherwise daughter cells would receive partly replicated DNA. A second control point is that S phase only occurs once in every cell cycle; once S phase has been completed it cannot be re-initiated until mitosis and cell division have occurred. If this was not the case, cell division would involve amplification of DNA content from generation to generation. There are, in fact, specialised cells in which DNA amplification does occur, one example being cells in the placenta derived from the fetus. This indicates that, under special circumstances, control points can be switched off.

3 The engine also exhibits checkpoints. These are points in the cell cycle where progress can be halted in the presence of adverse conditions. A good example of a checkpoint is the consequence of damaging DNA by exposing cells to ionising radiation. When DNA damage occurs, entry into S phase is stalled and re-initiated once the damaged DNA has been repaired. The machinery that comprises checkpoints is therefore normally silent but revealed by stressing cells in certain ways. An alternative way of looking at checkpoints is to say that certain conditions must be met (e.g. DNA is undamaged) before landmark events of the cell cycle can occur. The consequences of replicating damaged DNA are, as we shall see in Chapter 8, highly undesirable.

4 As already stated, the engine exhibits conservation of mechanism. In other words, all cell cycles are fundamentally similar. What this means is that, for example, components of the mammalian engine can be cloned by virtue of their ability to execute the same function in yeast. It also means that exotic cell cycles can be very illuminating about normal cell cycles. In this chapter we shall therefore be frequently departing from the narrow vision of the 3T3 cell in culture.

Entry into mitosis

We shall begin our analysis of the cell cycle engine by looking at the mechanisms that control entry into mitosis. This may seem to be out of order but, as shall emerge, an understanding of mitosis provides key insights into the rest of the cell cycle engine.

A range of experimental evidence indicates that mitosis involves the activation of some process that is kept silent for the rest of the cell cycle. This may seem obvious but it is remarkable to contemplate how a genetic event occurring at the restriction point can, in mammalian cells, lead to the activation of a process some 18–20 hours later. The evidence, in the main, involves mixing the cytoplasmic contents of cells in mitosis with cells at other points in the cell cycle.

One of the earliest examples of this approach involved fusion of mammalian cells in culture. In this experiment cells undergoing mitosis were fused with cells in G1, S or G2 and the behaviour of the resulting hybrid cells was examined (Fig. 6.1). The outcome of these experiments is clear. Fusion of a mitotic cell with a cell at any other point in the cycle leads immediately to induction of mitosis in the non-mitotic partner nucleus. This indicates that mitosis involves a mechanism which can override whatever else is happening in the cell cycle, and therefore this process must be switched off or suppressed during the majority of the cycle.

A more detailed understanding of the underlying machinery comes from

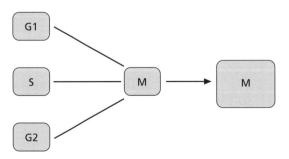

Fig. 6.1 Fusion of cells at different stages of the cell cycle. Fusion of mitotic cells with cells at any stage of the cell cycle results in immediate chromosome condensation, nuclear dissolution and mitotic spindle formation. This indicates that mitosis involves some activity which does not normally occur in the remainder of the cell cycle.

examining exotic cell cycles. The early post-fertilisation development of some species, such as amphibians, is characterised by very rapid and synchronous cell cycles. Combined with the large size and accessibility of amphibian eggs, this system has become a favourite preparation for examination of M phase control. The rapid cell cycles in early amphibian development arise from two causes: one is the firing of multiple DNA replication origins such that S phase is completed far faster than usual, and the other is an extreme contraction of the duration of G1 and G2. The net result is that cleaving amphibian embryos can, in effect, be considered as existing in one of two alternative states: S phase and mitosis.

Prior to fertilisation, the egg is held in a form of suspended animation before the completion of chromosome segregation and cell division of meiosis I. The egg can be induced to enter meiosis I (the first meiotic metaphase) by exposure to progesterone. This phenomenon is called 'maturation' and, for our purposes, can be considered analogous to a normal mitosis. If cytoplasm from an egg induced to undergo maturation by exposure to progesterone is injected into an immature oocyte, the recipient oocyte itself undergoes meiosis I. This implies that the induction of maturation by progesterone has activated a factor that is able to induce meiosis on a virgin nucleus in a similar manner to the cell fusion studies described above. This agent was termed maturation promoting factor (MPF). MPF has a remarkable 'self-catalytic' property: cytoplasm from an egg activated by injection of MPF can be injected into another immature egg and that will itself undergo mitosis and so on. In other words, MPF appears to be able, by some means, to induce MPF activity (Fig. 6.2) and this can be used to 'assay' MPF function.

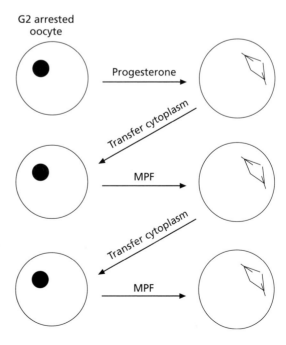

Fig. 6.2 The experimental basis of maturation promoting factor (MPF). G2 arrested frog oocytes can be induced to undergo mitosis in response to progesterone. The cytoplasm of mitotic oocytes will induce mitosis when transferred to a fresh G2 arrested oocyte. This MPF activity can be transferred to another oocyte.

The basic experimental design of the MPF experiment can be used to ask more questions about the underlying mechanism. MPF activity is exquisitely sensitive to inhibition of protein synthesis. This suggests that whatever MPF is, it involves an unstable protein that is synthesized *de novo* upon maturation. Surprisingly, however, MPF does not require gene expression or transcription. This can be demonstrated by maturing an enucleated egg with progesterone and finding the presence of MPF activity as usual. This indicates that MPF activity is due to a latent biochemical mechanism that is already present in the immature oocyte and which is activated by translation of an unstable protein at the onset of meiosis I.

Is MPF activity involved in a normal mitotic cell cycle? This can be tested by taking advantage of the synchroncity of the early cleavage divisions of amphibians (and indeed of other organisms such as echinoderms). Cytoplasm from S-phase and M-phase embryos can be injected into immature oocytes and tested for the presence of MPF activity. The results of these experiments are very clear. MPF activity rises and falls during cleavage-stage cell cycles in exact synchrony with the onset of mitosis (Fig. 6.3). This type of experiment can be carried further: all M phase cells that have been tested, including mammalian cells, exhibit MPF activity. The molecular mechanism manifest as MPF must be able to act

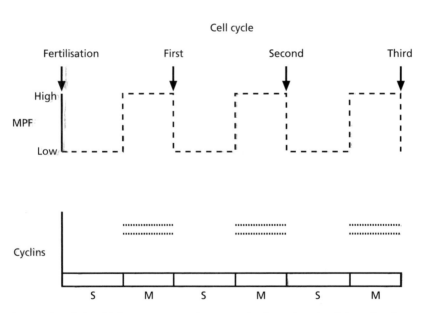

Fig. 6.3 The relationship between maturation promoting factor (MPF) activity, cell cycle phases and cyclins. The early synchronous cell cycles that follow fertilisation of amphibian eggs involve alternating M and S phases. MPF activity (see Fig. 6.2) abruptly increases at the onset of M phase and decreases at the end of M phase. The presence of MPF activity is closely linked to the appearance and disappearance of cyclins, as measured by gel electrophoresis of newly synthesized proteins.

over a very wide range of species, and therefore presumably exhibits strong conservation of function.

The appearance and disappearance of MPF activity in normal cell cycles coincident with mitosis means that we can redefine the cell cycle as having two components: 'on', in which MPF is active, and 'off', when MPF is inactive. When MPF is 'on', the cell enters mitosis. For the cell to exit mitosis, MPF must be destroyed in order to switch it 'off'. Putting all these pieces of experimental evidence together, we are led to the conclusion that MPF activity must involve the periodic translation and destruction of specific protein(s).

Is there any evidence for the existence of proteins which specifically appear at the onset of mitosis and disappear at the end? This can be tested by simply labelling synchronously cleaving early embryos with radioactively labelled amino acids and looking for labelled protein species that appear and disappear on schedule. Indeed, when this type of experiment is performed, one finds exactly what is expected: a family of proteins of around 35 kDa whose appearance oscillates in synchrony with mitosis. These proteins are called, by virtue of their cyclical fashion of expression, cyclins.

The discovery of cyclins raises a number of important questions. What is the relationship between cyclins and MPF? Are cyclins involved in activating MPF? What controls the destruction of cyclins? How does MPF act to coordinate mitosis?

To answer these questions we must first go to an entirely separate part of the evolutionary forest.

The genetics of cell cycle control in yeast

Yeasts are extremely useful organisms for genetic analysis of biological problems by virtue of their small size, rapid growth, small genomes and ease of physiological manipulation. We have so far ignored yeasts entirely in our account for a very good reason. They are unicellular organisms whose multiplication is controlled by two elemental forces: food and sex. They would not therefore be expected to have many of the control systems needed for the more complicated multicellular lifestyle. However, the cell cycle engine is precisely a problem for which yeast is well suited as an experimental organism. Yeasts obviously possess an engine mechanism, otherwise they would not divide. Their biological properties make it very easy to isolate and analyse mutations which affect cell cycle control. For this reason we shall now examine what has been learned from genetic analysis of the cell cycle in yeast.

Schizosaccharomyces pombe (*S. pombe*) is a yeast which divides by fission: during the cell cycle the cell increases in size, and at mitosis the cells divide into two equally sized daughter cells which then grow in size until the next mitosis. This property means that it is easy to score the stage of the cell cycle by the size (length) of the cell (Fig. 6.4). It is thus possible to isolate various types of (temperature-sensitive) *S. pombe* mutants that undergo mitosis at the wrong

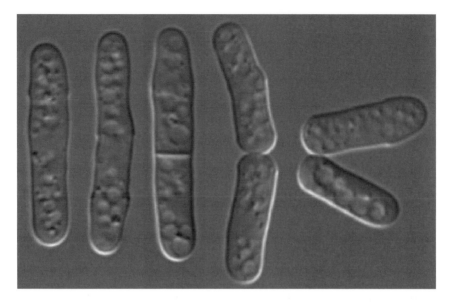

Fig. 6.4 The *Schizosaccharomyces pombe* cell cycle. Cells increase by growth at their tips until a critcal size is reached, whereupon the cells can enter mitosis and divide to form two daughter cells that are half the size of the parent prior to mitosis. (Courtesy of http://www.bio.uva.nl/pombe/cycle/dipl.gif)

size. These fall into two general classes: CDC mutants continue to elongate at the restrictive temperature and do not enter mitosis—they are therefore larger than wild type. The wee mutants, on the other hand, enter mitosis at the restrictive temperature at a smaller than normal size. CDC mutants are blocked in the entry to mitosis and wee mutants enter mitosis prematurely.

An especially informative *S. pombe* CDC mutant is CDC2. Most CDC2 mutants isolated are loss-of-function mutants which exhibit the characteristic elongated phenotype at the restrictive temperature. However, a mutant with the wee phenotype was isolated and proved to be a dominant-acting mutation in the *CDC2* gene. CDC2 therefore regulates the entry into mitosis. 'Inactive' CDC2 prevents entry into mitosis and 'activated' CDC2 accelerates entry into mitosis.

What sort of protein is encoded by the *CDC2* gene? It is a protein kinase of mass 34 kDa. It would therefore be reasonable to suppose that the control of mitosis exerted by *CDC2* is related in some way to the activity of its kinase. Indeed, this is the case since mutations in the *CDC2* gene which inactivate kinase activity result in the loss-of-function elongated phenotype. An unexpected feature of *CDC2* gene expression, considering its periodic function, is that the protein is expressed at virtually the same levels throughout the cell cycle. This means that the function(s) of *CDC2* must be due to periodic activation of CDC2 kinase activity by some other factor rather than simple regulation of the cellular levels of the CDC2 protein.

A truly remarkable feature of CDC2 is that its functions (and sequence) are highly conserved. It proved possible to isolate a CDC2 homologue from human cells by virtue of its ability to complement CDC2 mutations in *S. pombe*. CDC2 therefore acts in essentially the same way in human cells as it does in yeast. In other words, CDC2 is a universal regulator of mitosis in all eukaryotes. This discovery proves to be a key step in understanding the molecular mechanism of the cell cycle engine. It says that the activities of the engine reflect the activity of the CDC2 protein kinase.

What is the relationship between CDC2, cyclins and MPF activity? This was revealed by two convergent strategies. Purification of MPF from frog eggs revealed that it is a protein complex composed of two subunits: one is the frog CDC2 kinase and the other is a cyclin. Molecular cloning of another CDC mutation in *S. pombe*, CDC13, revealed that it encodes a cyclin. Cyclins are therefore required for mitosis and physically associate with the CDC2 kinase. These findings point to the possibility that CDC2 kinase activity, and consequent mitosis-inducing functions, are due to activation of CDC2 by association with cyclins. Destruction of cyclins at the end of M phase switches off the CDC2 kinase activity. That this is indeed the case was proved by two complementary experiments. In the first, an egg extract is treated with RNAse to remove endogenous, latent, cyclin mRNA. This extract is unable to induce mitosis in a heterologous nucleus until exogenous cyclin mRNA is added to the extract. This experiment shows that mitosis is dependent upon the presence of functional cyclin.

The second experiment is quite clever. Sequence analysis and mutation of cyclins revealed a region of conserved amino acid sequence required for proteolytic degradation. This is the so-called destruction box which targets proteins for conjugation with ubiquitin and consequent proteolytic degradation. The addition of mutant cyclins lacking the destruction box to egg extracts results in induction of mitosis, but the nuclei arrest and are unable to exit mitosis in the normal fashion. Cyclins are therefore required to enter mitosis and destruction of cyclins is required to exit mitosis. The MPF cycle described above is therefore essentially a cyclin/CDC2 cycle.

How do cyclins activate CDC2 kinase activity? Considerable insight has come from determination of the three-dimensional structure of a cyclin/CDC complex (Fig. 6.5). As might be expected, CDC2 is a typical kinase with the characteristic bilobed structure, in which the N-terminal domain contains the ATP-binding site and the C-terminal domain bears the active site. In between these two domains runs the activation loop, which has to be moved for the two kinase domains to interact in the act of catalysis. The binding of cyclin to CDC2 causes a subtle shift in the orientation of the activation loop and this brings the catalytic domain and the ATP-binding domain into close proximity, thereby allowing the kinase to be catalytically active. In the absence of cyclins the kinase domain is physically separated by the presence of the activation loop.

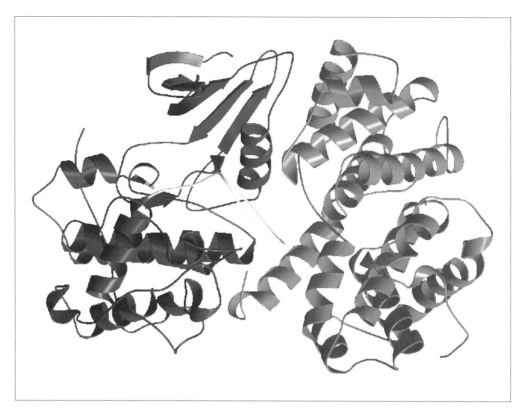

Fig. 6.5 The three-dimensional structure of the complex between cyclin B and the cyclin-dependent protein kinase, CDC2. The interaction of cyclin with the kinase induces relocation of the inactivation loop (lighter colour), allowing access of substrates to the substrate-binding domain and concomitant catalysis.

Other cell cycle mutants

What are the functions of all the other CDC and wee mutants of *S. pombe*? In particular, what are the biochemical functions of wee mutants that act to accelerate entry into mitosis? The identity of these genes might be expected to throw light on other potential mechanisms of CDC2 kinase activation.

Although tyrosine kinases are abundant regulatory molecules in multicellular organisms, tyrosine phosphorylation was found to be very rare in yeast. An important exception is CDC2, which was found to be phosphorylated on Tyr15 in interphase cells. Phosphorylation of Tyr15 inhibits CDC2 kinase activity. An obvious candidate for the inhibitory kinase that phosphorylates Tyr15 would be a gene encoded by one of the wee mutants and, indeed, *wee1* of *S. pombe* proved to encode just such an enzyme. Loss of wee1 function leads to failure of CDC2 phosphorylation and accelerated entry into mitosis. If tyrosine phosphorylation inactivates CDC kinase activity, it seems likely that this must be removed

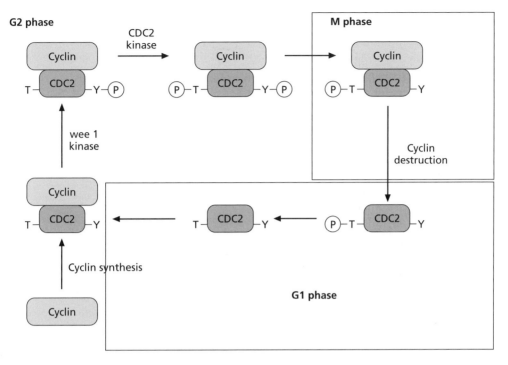

Fig. 6.6 The 'cyclin cycle' of *Schizosaccharomyces pombe*. In the G1 phase, CDC2 is dephosphorylated. Prior to mitosis, cyclin protein is produced and complexed to the receptor. CDC2 is phosphorylated on tyrosine 15 by wee kinase, which blocks kinase activity, and on threonine 161, which is required for kinase activity. Tyr 15 is dephosphorylated by the CDC25 phosphorylase and this permits activation of the cyclin/CDK kinase activity.

before the CDC2/cyclin complex can be active. It would be predicted on these grounds that the CDC2 phosphatase should exhibit a CDC phenotype. CDC25 of *S. pombe* encodes the phosphatase that dephosphorylates Tyr15 of CDC2.

A second important phosphorylation site on CDC2 is Thr161, which lies in the activation loop of the CDC protein (Fig. 6.5). Thr161 has exactly the opposite role to Tyr15: Thr 161 phosphorylation is required for CDC2/cyclin activity. This presumably involves displacement of the activation loop. The kinase that phosphorylates this residue in *S. pombe* is called CAK and, exceptionally, exhibits little homology to the equivalent CDC2 kinases in human cells.

Taken together, the picture emerges that CDC2 activity in yeast (and humans) is controlled by several different controls acting in concert (Fig. 6.6). One level of control is the requirement for cyclins to be bound for CDC to be active. The second is that, even in the presence of cyclins, at least two different and opposing phosphorylation events are required for activity: (i) the removal of an inhibitory phosphate; and (ii) the addition of an activating phosphate. This multiple control system surely provides a molecular example of an appropriate control system for ordering events in the cell cycle in that several independent events are required before kinase activity is unleashed.

The G1 to S phase transition

The analysis of the G2 to M phase transition described above has uncovered a key role for the cyclin/CDC enzyme complex in the induction of one landmark event. Is this is a special mechanism or is it possible that the cyclin/CDC system might be a universal regulator of all cell cycle landmarks? In order to answer this question we shall turn to the analysis of genes that control the transition from G1 to S phase in yeast.

Saccharomyces cerevisiae (*S. cerevisiae*) is probably the most popular strain of yeast for genetic analysis and, indeed, is the first eukaryotic organism for which all genes are known and sequenced. Although a unicellular fungi like *S. pombe*, *S. cerevisiae* is in fact a very distant relative of *S. pombe* and exhibits a very distinct lifestyle. *S. cerevisiae* is a budding yeast: rather than growing in size and splitting into two half-sized daughters, *S. cerevisiae* grows asymmetrically by budding (Fig. 6.7). In the presence of appropriate nutrients, the cell produces a small bud that grows progressively in size through the cell cycle. Mitosis occurs when the bud has reached approximately the same size as the mother cell. It is possible to stage the position in the cell cycle by examining the size of the bud.

A second important feature of the life cycle of *S. cerevisiae* is that the decision

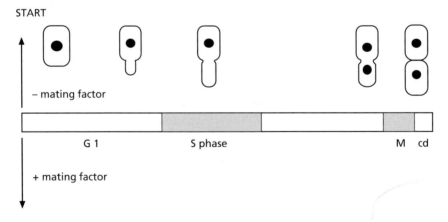

Fig. 6.7 The cell cycle of the budding yeast *Saccharomyces cerevisiae*. Prior to START, a cell has two options: in the presence of mating factor the cell enters sporulation and does not divide. In the absence of mating factor and the presence of nutrients, the cell passes START. This is manifested by the presence of the bud, which grows in size throughout the rest of the cell cycle. Mitosis is initiated as the bud reaches a similar size to the mother cell, and two daughter cells are produced.

to complete the cell cycle is controlled by environmental conditions. At the beginning of G1 a cell has two alternative fates: in the presence of mating factors (specific soluble peptides), the cells commit to enter mitosis. In the absence of mating factor and in the presence of nutrients, the cells commit to exit G1 and activate the cell cycle engine. The point in the cell cycle at which *S. cerevisiae* cells commit to activating the engine is called START. It will be seen that the phenomenon of START exhibits at least a superficial resemblance to the restriction point in mammalian cells.

As with *S. pombe* (which exhibits a very similar START feature), this biological feature of *S. cerevisiae* provides an excellent starting point to isolate (temperature-sensitive) mutations in genes that are involved in cell cycle control. In principle, it is possible to imagine two classes of mutant phenotype. In one case, the mutants would exhibit variability in the size of buds, indicating that they had arrested a variable time after START. In the second case, mutants would have uniform (or absent) buds which indicate that they had arrested at a unique point in the cell cycle. This basic idea can be elaborated to rank mutations in terms of the relative order in which they are required in the cell cycle.

Using this approach a large number of CDC mutants were isolated which arrested at different phases of the *S. cerevisiae* cell cycle. An especially interesting mutant was CDC28. CDC28 mutations have the unusual property of acting at two different stages of the cell cycle. In the first case, CDC28 mutant cells were defective in inducing cells to undergo START. This suggests that the CDC28 gene product was somehow involved in the mechanisms of START. On the other hand, by taking advantage of the temperature-sensitive phenotype, CDC28 mutant cells were shown to be also defective in their ability to undergo mitosis. In this respect they were analogous to the CDC mutants of *S. pombe*.

What is encoded by the *CDC28* gene of *S. cerevisiae*? It turns out to be none other than the cerevisiae homologue of the *S. pombe* gene *CDC2*. Indeed, it is possible to rescue CDC2 mutants of *S. pombe* with the homologous *CDC28* gene of *S. cerevisiae* and vice versa. This reveals that the CDC2/28 protein kinase has not one but two essential regulatory functions: one at the G2 to M transition and another at the G1 to S transition (or START).

How can one gene have two different functions? This is especially interesting in that, as described above, a reasonable amount is known about the control of CDC2/28 kinase activity in the G2 to M transition. Here again, the genetic approach proves informative. Mutants which are defective in their ability to respond to mating factor, and therefore START in the presence of mating factor, may be one of two types: the majority are defective in some aspect of the cell's response to mating factor. In a small minority, the phenotype arises from dominant-acting mutations which result in constitutive activation of START. Characterisation of three such genes, called *Cln1*, *Cln2* and *Cln3*, reveals that they encode cyclins that are distantly related in sequence to the M phase cyclins of *S. pombe*. These three new cyclins have been called G1 cyclins since they execute their functions in the G1 to S phase transition by association with the CDC2/28 protein kinase. Surprisingly, it appears as though the three G1 cyclins of *S. cerevisiae* have very similar functions since it is necessary to genetically in-activate all three genes before any strong effect on cell multiplication is evident. Conversely, it is clear that over-expression of G1 cyclins can accelerate entry into S phase and, significantly, rescue certain types of CDC2/28 mutations.

This genetic analysis has revealed that the CDC2/28 protein kinase/cyclin system is a universal regulator of cell control points. By combining different cyclins with a common kinase subunit, different cell cycle control points are traversed. Or, to put it a different way, the cell cycle engine is a 'cyclin' engine: the ordering of events is controlled by the ordered expression of different cyclin subunits.

Stating the problem in this way raises the questions: what controls the ap-pearance and disappearance of different types of cyclins? In the case of the G1 cyclins the answer is that their expression is controlled at the level of transcrip-tion. Cln1 and Cln2 transcription is induced abruptly at START. G1 cyclin ex-pression is also suppressed by the action of the G2 to S phase CDC2/28 complex. This provides a mechanism to ensure that the G1 and G2 type cyclins cannot be expressed concurrently. Thus, in order for the cell to progress from G1 to S phase at least two separate events must occur: destruction of the G2/M phase cyclins (by ubiquitin-mediated proteolysis); and activation of G1 cyclin transcription. This provides, in part, a means for the cell to 'know' whether it is in the G1 or G2 phase of the cell cycle, since if a cell is in G2 it actively suppresses the expression of cyclins that are typical of G1.

Cell size

The major value of yeasts in the study of the cell cycle engine is the relationship between the size of the cells and their position in the cell cycle. However, the account of the cell cycle engine described above does not actually explain how the relationship between cell size and cell cycle progression is controlled. The complete answer to this issue is not known, but one feature is that, in some cases, cyclin availability is also linked to net protein synthesis or, more specifi-cally, the content of ribosomes involved in protein translation. The G1 cyclin Cln3 is extremely unstable and levels of mRNA vary little during the cell cycle. The Cln3 mRNA has a 5' leader sequence in its mRNA which confers very low translational efficiency on the message. This means that the production of Cln3 protein is highly dependent upon the ribosome content of the cell: when the ribosome content of the cell is low, as, for example, in the presence of nutri-tional starvation, Cln3 protein is produced in very low amounts and cannot trigger cell cycle progress. However, in the presence of increased nutrients, ribosome content and general protein synthesis increase. This leads to en-hanced production of Cln3 protein until some threshold amount of protein is produced and kinase activation can proceed. The behaviour of Cln3 may represent a more general mechanism of linking cell growth and protein biosynthesis to cell cycle progression.

Mammalian cells

We know that the G2 to M phase transition in mammalian cells is controlled by an essentially similar mechanism to that in yeasts. Does this also hold true for the G1 to S phase control? And is this control mechanism in any way related to the restriction point?

As a result of conservation of both sequence and function, it has proved possible to clone mammalian homologues of both cyclins and cyclin-dependent kinases (CDKs). This has revealed that, unlike the situation in fungi, multiple families of both cyclins and CDKs exist. We can be fairly sure, having encountered this phenomenon several times already, that this elaboration of gene families is a means to create sophisticated levels of control by combinatorial associations of cyclins and CDKs. These can be broadly characterised by the presence of sequences (the PSTAIRE motif) involved in cyclin/CDK interaction (Fig. 6.5). The different cyclin and CDK families within mammals act in specific combinations. For example, the D-type cyclins largely associate with CDK4 and CDK6, whereas the E-type cyclins interact with CDK2.

A key feature of the different cyclin families in mammalian cells is that they exhibit different patterns of expression. For example, the E-type cyclins are expressed at the G1/S phase boundary, which suggests a function of the cyclin E/CDK2 complex in the firing of DNA replication. Most importantly, the D-type cyclins are expressed in the middle of the G1 phase of the cell cycle in response to growth factor stimulation (Fig. 6.8). This behaviour is exactly that of the intermediate class of growth factor-inducible genes, amongst whose members are predicted to be genes involved in the phenomenon of the restriction point. Can it be possible that it is the transcriptional induction of the D-type cyclins in G1 phase that starts the mammalian cell cycle engine?

The correct answer to this question is 'perhaps'. We will come back to the functions of mammalian G1 cyclins in Chapter 8. There is, however, very good evidence that D-type cyclin/CDK4 action can induce mammalian cells in culture to commit to DNA synthesis. Injection of preformed complexes between cyclin D1 and CDK4 into quiescent cells leads to the induction of DNA synthesis and bypasses the requirement for growth factor stimulation. A similar finding is seen by injection of preformed cyclinE/CDK2 complexes, but not with complexes of cyclin A and CDK2 which are normally active in S phase itself. This indicates that the different functions of cyclin/CDK complexes in mammalian cells reflect their temporal order of appearance, and that, in the G1 to S phase transition, cyclin D complexes act upstream of cyclin E complexes (Fig. 6.8). Moreover, injection of antibodies against cyclin D during G1, but no later, blocks the ability of serum-stimulated cells to enter DNA synthesis. Although this evidence seems persuasive, there is very good evidence that things are not

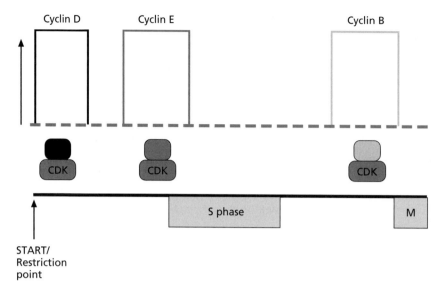

Fig. 6.8 Cyclin switching during the cell cycle. The cell cycle in mammalian cells involves an ordered activation of cyclin-dependent kinases (CDKs) which is orchestrated by the availability of specific cyclins at different points in the cell cycle. D-type cyclins are expressed prior to S phase at the restriction point. E-type cyclins are made available at the onset of S phase and B-type cyclins at the onset of mitosis.

so simple. The 'hard' model of mammalian G1 cyclin function, based upon the foregoing data, would suggest that an animal lacking cyclin D would fail to develop because its cells would be unable to get through G1. In fact, mice homozygous for inactive cyclin D1 genes are relatively normal and show defects only in the development of the retina and teeth. In addition, fibroblast cells derived from these animals exhibit normal growth characteristics and serum responses in culture. The provisional conclusion must be either that the presence of multiple cyclin families indicates they are able to compensate for each other's function or that things are more complicated than they seem.

Conclusions

The cyclin/CDK system is a highly conserved and biochemically beautiful mechanism that drives the major control points of the cell cycle engine in all eukaryotic organisms. Indeed, regulation of cyclin/CDK activity can account for all the central features of the cell cycle engine defined above. Different control points are defined by the combination of cyclin/CDK complexes in action. The activity of cyclin/CDK complexes accordingly represent switch points or milestones in the cell cycle. As we shall see later, cyclin/CDK complexes also provide a potentially important target for checkpoint controls if mechanisms that sense cell damage are linked into cyclin/CDK action.

Despite the persuasive attraction of this system, some important questions remain. Firstly, it is clear that the ordering of events in the cell cycle must involve some kind of interaction between cyclin/CDK complexes which prohibit concurrent action. The details of these controls are only partly understood. Secondly, it is clear from the study of mitotic cyclin/CDK complexes in *S. pombe* that additional levels of control, involving both positive and negative phosphorylation events, also play a key role in controlling CDK kinase activity. The importance of this type of control in mammalian cells and the identity of the relevant kinases and phosphatases is only beginning to be understood. We still do not understand how the activity of cyclin/CDK complexes controls the duration of different phases. Finally, we have not yet explained how the kinase activity of CDKs elicits coordinated changes in cell function. We shall return to this area in Chapter 8.

Chapter 7: Oncogenes

Introduction

Our attention so far has been centred on the 'normal' cell cycles of normal cells. In this chapter we shall examine what the cell cycles of 'abnormal' cells can reveal about biochemical mechanisms of cell proliferation. The abnormal cells in question are transformed cells and the biochemical mechanisms to be revealed are oncogenes.

Oncogenes are essentially genes which are experimentally defined by their ability to elicit the transformed phenotype in cells in culture. Transformed cells, as described in Chapter 1, are cells which are able to multiply with a greatly reduced or non-existent requirement for exogenous growth factors. These are both very simplistic definitions, but they are practically useful for our purposes since oncogenes represent pieces of the biochemical machinery of cell multiplication. In order for a gene to have the functional property of replacing signals from the cell surface, it must have some specific regulatory interaction with, or role in, the normal processes of cell proliferation. Hence, in identifying an oncogene, we have found a potentially important part of the biochemical mechanism of cell multiplication. The challenge is to define exactly what role the part plays in the functions of the normal cell.

We shall also learn that, in many cases, in order for a gene to be an oncogene it has to be either expressed in some specific fashion or altered in some specific way. The exact requirements vary from gene to gene but the nature of these alterations also potentially discloses significant information about the normal function of the gene in question. As we shall discover, the products of many oncogenes are, in fact, molecules that we have encountered already in previous chapters. Others, however, represent novel aspects of the cell cycle machinery which would be most unlikely to have been easily discovered by other means.

Oncogenes also have enormous practical value. Transformed cells are tumorigenic and therefore, by definition, oncogenes potentially play a role in human cancer. Moreover, many oncogenes were actually identified from the study of genetic changes in naturally occurring or induced tumours of humans and animals. Together, this means that understanding the identity and function of oncogenes has an important contribution to make to the diagnosis and therapy of human tumours.

The origins of oncogenes

There are two sources of oncogenes. Certain infectious DNA viruses have the property of either transforming cells in culture or being closely associated with, or the cause of, naturally occurring tumours. These viruses encode genes which have oncogenic functions. Such virally encoded oncogenes have no host cell counterparts and are required by the virus for some aspect of its infectious life-cycle. The oncogenic properties arise as a consequence of the interaction between the viral gene product and the host cell cycle machinery. Examples of such genes include the *Large T* gene of the SV40 virus, the *EIA* gene of tumour-associated adenoviruses and the *E6* gene of human papilloma viruses. We shall return to this set of genes in the next chapter.

The most prolific source of oncogenes is modified versions of normal cellular genes. Any normal cellular gene which can be altered in either sequence or expression to elicit cell proliferation in the absence of growth factors is potentially an oncogene. Conversely, the isolation of a modified cellular gene that confers the transformed phenotype has a corresponding normal counterpart which is, in principle, involved in the control of cell multiplication. This leads to the concept of a 'proto-oncogene', which is the normal cellular counterpart of an oncogene.

How are oncogenes identified?

The first examples of oncogenic genes were discovered in association with another class of viruses: the retroviruses. These viruses have an unusual lifestyle (Fig. 7.1). In their infectious state they comprise an RNA genome wrapped in a coat protein. When the virus infects a cell the RNA genome is copied into double-stranded DNA by virally encoded reverse transcriptase, and the DNA copy of the viral genome integrates into the host genome. Under normal circumstances, transcription of the integrated viral genome results in the production of a new infectious virus.

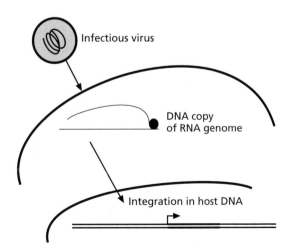

Fig. 7.1 The life cycle of retroviruses. The infectious virus enters the cell and its RNA genome is copied into DNA by reverse transcriptase encoded by the viral genome. The double-stranded DNA copy of the viral genome is then integrated into the host genome. Further infectious virus can be produced by transcription of the integrated viral genome.

Infectious virus

DNA copy of RNA genome

Integration in host DNA

Two different types of tumours are associated with retroviral infection. The first, more severe type, can be produced by infection of a host animal with the virus. The classic example is the Rous sarcoma virus which induces sarcomas (soft tissue tumours) upon infection of chickens. Other examples are listed in Table 7.1. Tumours induced by these viruses are frequently very aggressive and kill the host within weeks or months. These tumours are also multiclonal in origin. This means that the tumour is composed of many cells that have been independently infected by virus.

The reason this type of virus induces the tumour is because it has, at some point in its life history, recombined with the host genome and incorporated a normal host gene into its own genome. This rare event has several consequences. It brings the normal cellular gene under the control of the viral regulatory elements. It is also possible that, in the process of recombination, the host gene may be mutated in some way. Thus, as a result of the act of recombination with the host genome, the host proto-oncogene is expressed abnormally and may also be altered in function. The analysis of this relatively rare form of tumorigenic retrovirus leads directly to the discovery of a number of proto-oncogenes (Table 7.1) as the normal homologues of the genes that have recombined with the virus.

The second type of tumour associated with infectious retroviruses is somewhat different. In this case, the tumour typically arises relatively slowly and may not appear in all animals infected. Some examples are listed in Table 7.1. In some cases the effect of retroviral integration is inherited from generation to generation and reveals itself as a genetic predisposition to form tumours. A good example is the inherited propensity of certain mouse strains to form mammary tumours. This results from inheritance of an integrated copy of the mouse mammary tumour virus (MMTV). The tumours formed in these situations are all monoclonal in origin. This means that the tumour arises from a single infected founder cell. There must therefore be a requirement for an additional event before the oncogenic potential of the virus is unleashed.

The reason this type of virus causes a tumour is that it has integrated in the vicinity of a proto-oncogene and has, as a consequence, brought the expression of the proto-oncogene under the control of its own promoter elements. It is transcriptional activation of the host gene by the virus that leads to the formation of the tumour. This, in part, explains the long latency and monoclonality of these types of tumour. It is required that the virus integrates into a specific genomic location to elicit its tumour-forming effect. This location has to be in the vicinity of a proto-oncogene whose oncogenic potential can be activated by alteration in expression.

The next method of isolating oncogenes involves directly cloning them from genomic DNA by virtue of their oncogenic function. The idea behind the approach is as follows: If tumours are the result of genetic mutations which lead to conversion of proto-oncogene into an oncogene, it should be possible to

Table 7.1 A partial list of known oncogenes, indicating their source and biochemical function where known. (From the Online Mendelian Inheritance in Man database.)

Oncogene	Source	Function
Growth factors		
FGF3	Retroviral insertion	FGF family member
FGF4	Transfection	FGF family member
FGF8	Retroviral insertion	FGF family member
Gro1	Transfection	Chemokine
PDGFB	Retroviral recombination	PDGF family member
Wnt1	Retroviral insertion	Wnt family member
Receptors		
Axl	Transfection	Receptor tyrosine kinase
Erb-b	Retroviral recombination	EGF receptor
Neu	Transfection/transformation	EGF receptor
FGFR1	Translocation/fusion	FGF receptor
Fms	Retroviral recombination	CSF-1 receptor
Kit	Retroviral recombination	SCF receptor
Mas1	Transfection	7TM receptor
Mpl	Retroviral recombination	Tpo receptor
PDGFR-beta	Translocation/fusion	PDGF receptor
Ret	Positional cloning/transfection	BDNF receptor
Transducers		
Abl	Retroviral recombination	Cytoplasmic/nuclear tyrosine kinase
Akt	Retroviral recombination	Membrane Ser/Thr kinase
Cbl	Retroviral recombination	Membrane tyrosine kinase
Fes	Retroviral recombination	Membrane tyrosine kinase
Fyn	Related to Src	Membrane tyrosine kinase
Hck	Related to Src	Membrane tyrosine kinase
Lck	Related to Src	Membrane tyrosine kinase
Mel	Transfection	G protein
Mos	Retroviral recombination	Cytoplasmic Ser/Thr kinase
Raf	Retroviral recombination	Membrane Ser/Thr kinase
Ral	Retroviral recombination	Ras family G protein
Ras	Retroviral recombination	Membrane G protein
Src	Retroviral recombination	Membrane tyrosine kinase
Vav	Translocation/transfection	Guanidine nucleotide exchange factor
Yes	Retroviral recombination	Src family tyrosine kinase
Transcription factors		
Elk-1	Translocation	ETS family
Erg	Retroviral recombination	ETS family
Fos	Retroviral recombination	Partner to Jun
Gli	Amplification	Zinc finger protein
HKR3	Homology cloning/transfection	Related to Gli
Jun	Retroviral recombination	Partner to Fos
Maf	Retroviral recombination	Leucine zipper protein

(*cont. on p. 96*)

Table 7.1 *Continued*

Oncogene	Source	Function
Max	Homology cloning/transfection	Partner for Myc
Myb	Retroviral recombination	Haemopoietic transcription factor
Myc	Retroviral insertion/recombination Translocation	Transcription factor
Rel	Retroviral recombination	NFκB
Ski	Retroviral recombination	Binds MAD family transcription factors
Cell cycle engine		
Cyclin D	Amplification	Activates CDKs
MDM2	Amplification	Degrades p53

directly clone oncogenic DNA sequences from tumours by virtue of the ability to transform normal cells.

The basic experimental strategy is outlined in Fig. 7.2. DNA is extracted from a specific type of tumour and transfected into a transformation-sensitive target cell, such as 3T3. This gives rise to rare transformed colonies as a result of their acquiring a DNA segment with oncogenic function. Since, in all likelihood, these cells have taken up vast amounts of normal cellular DNA, this is 'diluted out' by taking DNA from the transformed colony and using it to retransform for a second round. This process is repeated until a minimal segment of transforming DNA containing the oncogene can be successfully cloned.

Some genes isolated by this approach are listed in Table 7.1. It should be noted that this approach does not prove that the gene isolated was the cause of the original tumour. There are a number of cases in which the gene that is eventually cloned turns out to have transforming activity as an artefactual deletion or mutations acquired during the transfection process. For example, a normal gene can integrate into host DNA near a powerful promoter or lose some negative regulatory elements in its promoter during the process of integration. This technique therefore identifies genes which can transform cells either as a result of mutation in the original tumour DNA or as a result of activation of expression in the process of transfection. A more subtle problem is that the type of genes that are isolated tend to reflect the type of cell which is used as the target for transformation. Using a single target, such as 3T3, tends to lead to repeated isolation of the same genes from a wide variety of tumour DNAs. This reflects the sensitivity of the target cell to the activity of the oncogene.

A common feature of many types of tumour is that they exhibit abnormal chromosomes. In particular, chromosome translocations are commonly associated with particular types of tumour. This is a situation where one segment of a particular chromosome breaks off and joins with another (Fig. 7.3). What is in-

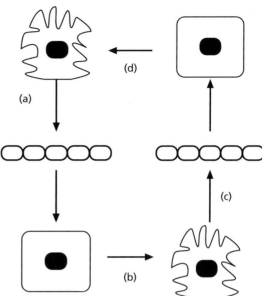

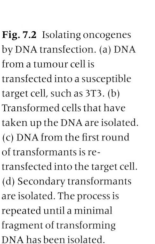

Fig. 7.2 Isolating oncogenes by DNA transfection. (a) DNA from a tumour cell is transfected into a susceptible target cell, such as 3T3. (b) Transformed cells that have taken up the DNA are isolated. (c) DNA from the first round of transformants is re-transfected into the target cell. (d) Secondary transformants are isolated. The process is repeated until a minimal fragment of transforming DNA has been isolated.

teresting about these cases is frequent associations between specific pairs of chromosomes in translocations associated with a particular type of tumour. Perhaps the best-known example is the Philadelphia chromosome, found in 95% of patients with chronic myelogenous leukaemia and which involves a translocation of a region of chromosome 11 of humans onto chromosome 22.

The underlying basis of these events is that, as a result of the translocation, a proto-oncogene has either been moved into the vicinity of a heterologous promoter or beeen fused with a protein-coding sequence to produce a hybrid protein. The characterisation of gene pairs involved in chromosomal rearrangements therefore leads to the identification of proto-oncogenes which can be activated either by ectopic expression or gene fusion (Table 7.1). It is interesting to

Fig. 7.3 Chromosome translocations. A fragment of one chromosome is broken and fuses with another chromosome. In some cases the translocation is reciprocal and two pieces from different chromosomes are interchanged.

note that most examples of gene fusions generated by this route result in oligomerisation and truncation of the proto-oncogene.

A final example of oncogene hunting also arises from another type of chromosomal abnormality in human tumours. This is the phenomenon of gene amplification. In this situation the tumour is found to have undergone local amplification of specific segments of DNA in which the target proto-oncogene lies. This results in the presence of a large number of copies of the gene in the transformed cell. The phenomenon of gene amplification reveals a very specific class of genes: those whose oncogenic functions are activated by being expressed at higher levels than their normal counterparts (Table 7.1).

What are oncogenes and what do they do?

The identification of oncogenes (and their proto-oncogene counterparts) essentially reveals a component of cell cycle machinery: a protein which, when subjected to oncogenic activation, has the ability to induce progress through the cell cycle. The principal interest from the perspective of cell cycle control is therefore to understand the biochemical function of the protein in question and how its oncogenic conversion leads to cell transformation.

A closer look at the sample of oncogenes identified by different techniques (Table 7.1) reveals that there is no obvious feature that these genes share apart from their ability to transform cells. Oncogenes (and proto-oncogenes) represent a diversity of functions, sequences and cellular locations. There is, in other words, no single explanation for cell transformation or tumorigenicity. However, this might be expected from what is already known of cell cycle control mechanisms. Following this line of thought it becomes clear that the majority of oncogenes fall into four familiar functional classes: growth factors, receptors, signal transducers and transcription factors. In this respect an important issue is the mechanism by which a proto-oncogene is turned into an oncogene. This will not only illuminate conditions under which oncogenic conversion events will occur but also reveal something about the normal function of the proto-oncogene, and hence its position on the cell cycle control map.

Growth factors

A number of oncogenes are growth factors. *sis* is the oncogene incorporated into the genome of the simian sarcoma virus and it encodes PDGF-B. The ability of the simian sarcoma virus to form tumours is therefore the result of expression of PDGF-B under the control of viral regulatory elements rather than its own promoter. A number of members of the FGF family—FGFs 3, 4 and 8—have been identified as targets for transcriptional activation by proviral insertion of the MMTV. In this case, integration of the viral genome in the vicinity of the FGF target gene results in expression of FGF under the control of the viral promoter in the mammary gland. A similar mechanism is involved in the activation of the

Wnt family of growth factors by MMTV proviral insertion. A final example is FGF4, which was isolated directly from genomic DNA by two routes: (i) transfection of DNA from Kaposi sarcoma tumours into 3T3 cells; and (ii) as a consequence of its inclusion in a region of DNA that is subject to DNA amplification in stomach tumours. In fact, it is mostly likely that in both these cases the isolation of *FGF4* as an oncogene is an artefact arising from the experimental protocol, rather than a reflection of the involvement of *FGF4* in the original tumour. The *FGF4* gene contains a number of strong repressor regulatory elements in its promoter which were separated from the *FGF4* coding sequence in the process of gene isolation. As a result, the transfected *FGF4* gene becomes transcriptionally activated and that is the basis of its ability to transform cells.

This illustrates a common principle in the oncogenic activation of growth factor genes: it almost invariably involves transcriptional activation of the gene, leading to production of the growth factor in an inappropriate place. In fact, this mechanism can be reproduced experimentally by forcing expression of growth factors in adult tissues by various means; in many cases the result is the production of a tumour in the cell type in which the heterologous growth factor is expressed. The ability of growth factors to transform cells therefore results from an autocrine mechanism (Chapter 1). The cell is no longer dependent upon external signals to induce cell proliferation since it produces its own.

There are two intriguing consequences of this mechanism of activation. First, it implies that, in many adult tissues at least, the reason cells are quiescent is because they are deprived of appropriate growth factor stimulation. Second, ectopic expression of a growth factor by a transformed or tumorigenic cell may stimulate neighbouring, non-transformed cells by a paracrine mechanism. This can lead to many of the secondary consequences of tumour formation such as fibrosis (the production of a mat of tissue, analogous to a scar around the tumour) or angiogenesis (the induction of new blood vessel formation in the vicinity of the tumour by stimulation of endothelial cells). These secondary consequences of ectopic growth factor expression are frequently significant in the 'natural history' of the tumour in the host.

Growth factor receptors

A significant number of oncogenes bear the defining hallmarks of growth factor receptors. They are transmembrane proteins with intrinsic tyrosine kinase domains in the cytoplasmic region.

The *neu* gene was isolated (as the name implies) from a chemically induced neuroblastoma in the rat by the DNA transfection route. The *neu* gene is in fact none other than HER-2, a member of the EGF family of receptors. However, the *neu* oncogene is subtly different from its normal counterpart—it has specific point mutations in the transmembrane domain. The result of these mutations is to induce the transmembrane domain of the HER-2 receptor to undergo spontaneous dimerisation.

A number of growth factor receptors have been identified in tumours as a result of chromosomal translocation events that lead to the production of fusions with other cellular proteins. Thus, both the *FGFR-1* and *PDGFRB* genes have been recovered from leukaemias as gene fusions with another gene, *TEL*. TEL is a transcription factor, and it may, at first sight, seem strange that fusion of a transcription factor with a growth factor receptor leads to oncogenic activation. Closer inspection of the protein products of these fusions provides the explanation. TEL is a dimeric protein and the oncogenic fusions invariably involve fusion of the dimerisation domain of TEL with the tyrosine kinase domain of the target receptor. This leads to forced dimerisation of the kinase domain and stimulation of kinase activity.

The essential mechanism of oncogenic activation of growth factor receptors is therefore forced dimerisation as a result of specific mutation or gene fusion. As a result, signal transduction pathways are activated without a requirement for growth factor stimulation. In this situation, the transformed cell acts as though it is receiving an extrinsic growth factor signal and begins to proliferate.

Signal transducers

One of the most frequently recovered classes of oncogenes are members of the Ras family of membrane-associated small G proteins (Chapter 4). Thus, members of the *Ras* gene family have been incorporated into the genomes of a number of different tumour-forming retroviruses such as the Harvey and Kirsten rat sarcoma viruses (hence the origin of the name Ras). *Ras* genes are also frequently recovered from tumorigenic DNA using the DNA transfection route. In all cases these oncogenic *Ras* genes are found to have undergone point mutations that directly or indirectly eliminate the intrinsic GTPase activity of the Ras protein. Indeed, point mutations of this type in *Ras* genes prove to occur frequently in naturally occurring human tumours of many types. The result of these mutations is that the *Ras* 'switch' is permanently switched to the 'on' position and is thereby uncoupled from a requirement for upstream activators. The oncogenic *Ras* genes therefore expose the cell to a continuous intracellular mitogenic signal.

Many intracellular protein kinases have the classical kinase design of catalytic and regulatory domains, in which the regulatory component acts to suppress kinase activity unless bound or modified by co-factors. Physiological de-repression of kinase activity can therefore be mimicked by mutations which abolish the inhibitory functions of the regulatory domain. This is a common mechanism of oncogenic activation of intracellular kinase transducers. Thus, for example, the *Raf* gene was originally identified as a transforming oncogene from the rat fibrosarcoma virus. Incorporation of the *Raf* gene into the viral genome resulted in truncation of the protein, resulting in constitutive activation of the kinase activity and concomitant activation of the downstream MAPK pathway.

Another family of kinases, whose position in the signalling cascade is less clear, is the membrane-associated cytoplasmic tyrosine kinases of the Src family. *Src* is the prototype member of this gene family and was the transforming gene incorporated into the prototype transforming retrovirus, the Rous sarcoma virus. Indeed, the *Src* gene has a special status in the history of oncogenes as it was also the first proto-oncogene to be identified on the basis of its incorporation into a transforming retrovirus. All oncogenic versions of Src have undergone truncations or point mutations that block the functions of an N-terminal inhibitory domain, thereby activating kinase activity.

In general, therefore, oncogenic signal transducers activate intracellular signalling pathways as a result of mutations which block or suppress regions of the protein involved in suppression or attenuation of activity.

Transcription factors

A considerable number of transcription factors have been found to have oncogenic function. Thus, the *fos* gene was identified as a transforming oncogene of the FBJ osteosarcoma virus and NFκB was identified as the transforming gene of the reticuloendotheliosis retrovirus of turkeys. In all cases, the mechanism of oncogenic activation is the introduction of mutations which constitutively activate the actions of the protein, leading again to uncoupling of activity from upstream activators.

In the case of the *fos* gene the mechanism of activation appears to result from truncations of the coding and non-coding sequences of the gene when it was incorporated into the virus. At the 3′ end of the gene a deletion was introduced which has at least two consequences. One is to remove the 'destability' sequences in the 3′ non-coding region of the gene. This results in stabilisation of the oncogenic *fos* message. The second is to remove a fragment of the C-terminal region of the normal protein which appears to be involved in transcriptional suppression of *fos* gene expression. The combined result of these changes is to produce a form of the protein which is expressed stably, rather than transiently during the normal cell cycle. In the case of NFκB (Chapter 4), the oncogenic form of the protein has been truncated at the N-terminus and this blocks the association with inhibitor IKB. This results in permanent localisation of the transcription factor in the nucleus and activation of downstream target genes without a requirement for activation by signal-mediated release from the inactive cytoplasmic complex.

A very intriguing example of an oncogenic transcription factor is *c-myc*. This gene has been frequently recovered as an oncogene by various routes: either directly incorporated in a transforming retroviral genome, located in the vicinity of a proviral insertion in a tumour, or subjected to chromosomal rearrangements. The latter event is the underlying basis of the 'Philadelphia' chromosome described above and which results in relocation of the *myc* gene to the immunoglobulin gene cluster. The result of this relocation is to bring the *myc*

gene under the transcriptional control of the powerful B-cell-specific immunoglobulin gene regulatory elements. In all these manifestations of the
oncogenic actions of the *myc* gene, the effects arise from elevated expression of
the normal gene rather than mutation and loss of function. This is significant
because, under normal circumstances of cell cycle progression, it appears as
though the *myc* gene is expressed transiently and at very low levels, implying
either that it is not normally active or that its actions are executed at limiting
amounts of protein.

It is clear therefore that the *myc* gene has some role in cell cycle progression
but what is this role? The normal cellular myc protein is a transcription factor of
the HLH box type, is located in the nucleus and (as shown by mutation of the
DNA-binding domain) transforms cells as a result of its interaction with DNA.
Presumably, therefore, its actions involve activation (or suppression) of downstream target gene expression. A defining feature of the HLH box family of transcription factors is that they form dimers with partner HLH box proteins as part
of their mechanism of action. Myc is indeed able to dimerise with other partner
proteins. Max is a dimerising partner of the normal *myc* gene product and this
partnership is required for myc to execute its transforming functions. Deletion
of the dimerisation domain of myc blocks the ability to transform cells. However, the situation is complicated by the fact that Max is also able to partner with
two other HLH box class proteins, Mad and Mxi1. Formation of these partner-
ships leads to sequestration of Max and inhibition of the transcriptionally com-
petent Myc/Max partnership. Mad and Mxi1 are expressed at high levels in
non-proliferating cells and therefore inhibit myc activity. When cells are induced to proliferate, the levels of Mad and Mxi1 decline and the productive
Myc/Max partnership can form. On this basis, the action of Myc seems to be a
manifestation of progress through the cell cycle rather than a cause. However,
this scheme provides an explanation for the oncogenic actions of the Myc protein. High-level expression of myc relieves the inhibition of Max by titrating out
the inhibitory partners, thereby escaping from normal control.

Myc provides an excellent example of the relationship between conven-
tional biochemical dissection of cell cycle control mechanisms and the oncogene approach. The fact that *myc* is an oncogene implies that it must be involved
in cell cycle control at some level, but the downstream targets for *myc/max* activation and the upstream signals that regulate *myc* functions are still obscure. Recent evidence indicates that one possible feature of Myc activity is the induction
of genes that are involved in the control of cell size—or more strictly protein
content (see Chapter 1). Forced expression of *myc* in B lymphocytes (as would
occur in cells harbouring an oncogenic *myc* gene) leads to an increase in cell size.
In addition, examination of the genetic targets of Myc activation reveals that
many of these genes encode proteins that are directly or indirectly involved in
cell metabolism and protein synthesis, such as translation initiation factors and
ribosomal proteins.

Cyclins (or other genes) which lie at the 'end' of the signalling cascade might, in theory, be obvious targets for oncogenic activation. Cyclin D is the only such target currently associated with a tumorigenic phenotype as it is found to be amplified in certain kinds of tumour and therefore expressed at higher levels than normal. In fact, further reflection will suggest that cyclins, and other components of the cell cycle engine, are in fact very poor targets for driving cells through the cell cycle since their essential functions depend upon their ordered appearance and destruction. Genetic changes which alter this pattern might be predicted to have catastrophic consequences for cell viability.

Oncogenes in perspective

In this chapter we have seen that the basic components of cell cycle control reappear in a different guise. Altering the function of a gene involved in the signalling cascades results in a capacity to uncouple cell proliferation from upstream signals, the production of proliferative signals and a cell which, in essence, proliferates when its normal counterparts do not. This is manifest in the ability to transform cells (especially established cell lines), which is, to the first approximation, equivalent to tumorigenesis. Indeed, many oncogenes have been identified by virtue of their intimate involvement in the formation of tumours in experimental animals or humans.

Oncogenes have, almost invariably, one of four generic functions: they can be growth factors, receptors, cytoplasmic signal transducers or transcription factors. Thus, the normal process of cell cycle control can be subverted at several different levels. It is important to note, however, that all cell cycle control mechanisms sensitive to oncogenic transformation lie upstream of the restriction point.

The exact mechanism of oncogenic transformation depends entirely on the biochemical function of the protein in question. But, in general terms, these changes either uncouple the activity of the protein from upstream signals or prevent inhibition of activity by other cellular proteins. In other words, oncogenes are essentially active when they should not be.

Finally, it is intriguing to contemplate the fact that all mechanisms of oncogene activation are essentially genetic in origin: they involve either coding changes to the gene itself or alterations in the transcriptional regulatory elements that influence expression of the gene. This might seem obvious but, as we shall see in the next chapter, this has profound consequences for our understanding of cell cycle control in malignant cells.

Chapter 8: Tumour Suppressor Genes

Introduction

So far we have taken the view that cell cycle control mechanisms are essentially 'active' in nature. That is to say, progress through the cell cycle involves the activation of a series of biochemical events in sequence. Each event is dependent upon prior activation of a previous event, leading back to the interaction of a growth factor with a cell surface receptor. This idea has been reinforced by the action of oncogenes where some type of mutation or altered gene regulation activates, at different levels, the normal mechanism of cell cycle control. This view is an oversimplification. In this chapter we shall deal with the identity and activity of genes whose actions are to prevent progress through the cell cycle unless certain conditions are met. These genes, by virtue of their most explicit activities, are called tumour suppressor genes.

The idea of oncogenic activation of a proto-oncogene by mutation or gene activation would predict that naturally occurring tumours would arise with 'single hit' kinetics. A single event—a mutation or a viral insertion—is sufficient to induce the formation of a tumour. This might be reinforced by the ability of a single oncogene to transform established cell lines such as 3T3. This would predict that an individual has an equal probability of developing a tumour throughout their lifetime. However, under most circumstances, it is perfectly plain that the process of tumorigenesis is more complex and exhibits 'multi-hit' kinetics. Thus, for example, it has been clearly established that, for many common types of cancer such as those of breast, lung and colon, the probability of developing a tumour increases with age. In cases where tumour induction can be ascribed to the action of a genotoxic (DNA-damaging) carcinogen, such as radiation or cigarette smoke, there is a relationship between the length of exposure to the carcinogen and the probability of a tumour appearing. Moreover, not all individuals exposed to a carcinogen go on to develop tumours, and there is frequently a long lag period between the time of exposure to the carcinogen and the appearance of a tumour. Taken together, this type of epidemiological evidence suggests that the formation of tumours is better explained by the requirement for a series of events to occur rather than a single event.

This concept is reinforced by experimental studies of activated oncogenes *in vivo*. For example, if an activated oncogene is expressed in all the cells of a particular tissue, the naive expectation might be that the resulting tumour would be multiclonal in origin, i.e. derived from all the cells in the tissue in which the oncogene was expressed. In reality, this type of experiment frequently gives a different result. The tumours that arise from expression of an activated oncogene are frequently monoclonal in nature (i.e. derived from a single founder cell) and arise after a relatively long lag period. This is reminiscent of the kinetics of tumour induction by retroviral insertion (Chapter 7) and clearly indicates that additional rare events, perhaps genetic in nature, need to occur before an activated oncogene can induce the formation of a tumour. Another way of looking at this would be to say that activation of a proto-oncogene is not sufficient to induce a tumour in a real animal and there must accordingly exist mechanisms in normal cells which inhibit the formation of tumours by activated oncogenes.

The existence of such mechanisms can be inferred from studies in cell cultures. Thus, if a normal cell (say a lymphocyte or a macrophage) is fused with a malignant cell (say a transformed 3T3 cell), the resultant hybrid, which contains the genomes of both partners, turns out to be non-malignant in behaviour. The genome of the normal cell therefore appears to contain genetic information which suppresses, by some means, the transforming activity of the genome of the malignant cell. However, if such hybrids are cultivated for a period in culture, it is frequently observed that chromosomal loss occurs and the genetic composition of the hybrid changes with time. When this happens, the malignant phenotype can reappear. Furthermore, the reappearance of the dormant malignant trait in hybrid cells can be correlated with loss of specific chromosomes, or regions of chromosomes, derived from the non-malignant parent (Fig. 8.1). This indicates that the suppression of the malignant phenotype in the hybrid cell is due to the presence of genetic information in the hybrid which is derived from the normal parent. In this type of experiment, therefore, the transformed phenotype exhibits 'recessive' genetic characteristics. This result, from the perspective of oncogene action described in Chapter 7, is somewhat counter-intuitive.

If the results of a large number of such experiments, involving widely differing malignant and non-malignant partners from a broad range of species, are considered together, it would appear that there exist multiple, but not many, regions of genome which have the ability to suppress malignancy in hybrid cells. This indicates that the ability to suppress the malignant phenotype in hybrid cells is the property of a relatively small number of genes that are relatively conserved in function between species.

Further indications that the phenomenon of cell transformation results from the collusion of multiple gene products comes from looking at the ability of oncogenes to transform primary, non-established, pre-senescent cells in culture. It is, in fact, remarkably difficult to transform primary cells by expression of a single activated oncogene compared with the relative ease of transformation observed with established cell lines. However, it is possible to transform primary cells with reasonable efficiency if pairs of activated oncogenes are employed. Testing pairs of oncogenes reveals that they fall into two classes or 'complementation' groups. In order to transform a primary cell, one member of each class is required. In one class ('immortalising oncogenes') are the products of

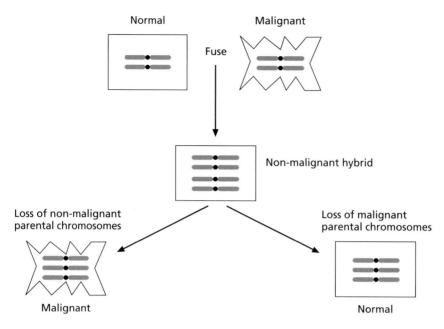

Fig. 8.1 Mapping malignancy in cell hybrids. Tumorigenic cells are fused to a normal non-malignant cell type. The resulting hybrid exhibits a non-malignant phenotype. As hybrid cells are propagated they undergo chromosome loss. If chromosomes from the malignant parent are lost, the cell retains the non-malignant phenotype. If chromosomes from the non-malignant parent are lost, the hybrid acquires the malignant phenotype.

DNA viral oncogenes such as SV40LT and adenovirus E1A and the myc oncogene. In the second class are the classic 'signalling' oncogenes derived by mutation of cellular proto-oncogenes such as *Ras* or *Src*. This finding has two important implications. Firstly, it suggests that the 'immortalising oncogene' class acts on biochemical pathways separate from those which activate mitogenic signalling processes. Secondly, it implies that the process of establishment and immortalisation which, it will be recalled, invariably involves chromosomal changes and genetic instability, can be mimicked by the action of 'immortalising oncogenes'. Taken together, these considerations suggest that 'immortalisation' or establishment results from loss or suppression of some cellular activity. This activity can be suppressed by the action of immortalising oncogenes derived from tumorigenic DNA viruses, and presumably is also inactivated by some means in naturally occurring tumours.

All the evidence described above clearly points to the existence of some biochemical mechanism(s) which prevents a cell from proliferating indefinitely in the presence of a continuous mitogenic signal. The hints from this evidence also indicate that this putative mechanism is in some way involved with genome integrity in that it can be inactivated or suppressed by destabilisation of the genome or mutation. What is lacking, however, is any clue as to the identity of the molecules involved, other than that they might interact with, or be regulated by, the actions of tumorigenic DNA viral genes.

A major advance in identifying these gene products comes from analysing the molecular basis of relatively rare genetic mutations. These are mutations that result in an inherited predisposition to form tumours. These mutations exhibit very high penetrance, i.e. virtually every individual who inherits the mutated gene will develop a tumour. They are also frequently tumours of childhood, which contrasts with the age dependancy of the majority of tumours in the human population. Finally, the unfortunate individuals who inherit these mutations frequently develop multiple tumours; this again contrasts with the majority of naturally occurring tumours.

Two examples of such tumours illustrate the point. Retinoblastoma is a relatively unusual childhood tumour of the retinoblast of the eye and exhibits a clear pattern of Mendelian inheritance. Individuals who inherit the retinoblastoma susceptibility gene frequently develop multiple retinoblastomas and also show a greatly increased propensity to develop other types of tumour in later life. Li–Fraumeni syndrome is a very rare inherited cancer susceptibility syndrome in which individuals who inherit the mutation develop multiple types of aggressive tumours in early childhood. In both cases the inherited cancer susceptibility syndrome is associated with specific chromosomal deletions, indicating that the basis of the inherited predisposition is the loss of specific regions of genetic information. It may be presumed that a gene located in the 'missing' region of DNA is involved in inhibiting the process of tumour formation. This consideration leads to the 'two-hit' hypothesis (Fig. 8.2), which suggests that, in order for a tumour to occur, both copies of a target gene must be inactivated by mutation. Thus, in normal individuals, two inactivating mutations, one in each copy of the gene, must occur. However, an individual who inherits a gene which has already undergone a loss of function mutation requires only a single additional mutation in the other copy of the gene for the tumour to arise. This would give rise to a situation where tumour formation was dependent upon a single genetic lesion, thereby explaining the rapid onset and highly penetrant phenotype. This hypothesis predicts that if the target gene could be identified, one copy would be mutated in the normal tissue and both copies will be mutated in the tumour. As we shall see below, this prediction has been amply confirmed.

Retinoblastoma and Li–Fraumeni syndromes are unusual, but clearcut, cases of the inheritance of mutated genes leading to tumour formation. How applicable is this idea to the majority of tumours where no obvious pattern of strict Mendelian inheritance can be discerned? A detailed examination of the clustering of specific types of tumour within families has led to the identification of additional genes which confer susceptibility to the formation of specific types of tumour. Two examples are *BRCA1* and *BRCA2* (for breast cancer associated); these are genes which are mutated in about 5% of women with early-onset breast cancer. It seems likely that mutations in these genes are not fully penetrant (i.e. not all individuals who inherit the mutation will develop a tumour)

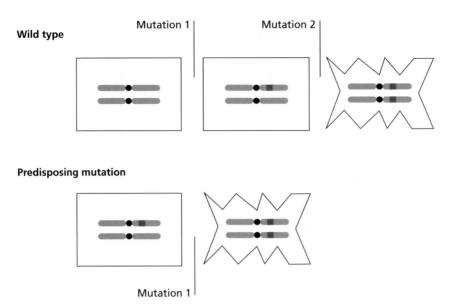

Fig. 8.2 The two-hit hypothesis of tumour formation. This supposes that both copies of a tumour suppressor gene have to be inactivated in order for a tumour to form. In the case of individuals who have inherited an inactive allele of the tumour suppressor gene, only a single further inactivating mutational event is required. This explains the rapid onset and multi-focal nature of tumours in individuals who inherit a mutated tumour suppressor gene such as *p53* or *Rb*.

and that, certainly, these two genes alone cannot account for the majority of breast tumours that occur in the population, implying that other genes remain to be identified. However, they illustrate the point that inactivating mutations in a crucial set of target genes may be responsible for the majority, if not all, of naturally occurring tumours.

Taking all the preceding evidence together, we are led to the idea that there are activities within the cell which must be inactivated by some means in order for a cell to proliferate indefinitely. These are encoded by tumour suppressor genes.

What is the identity of tumour suppressor genes and what do they do?

Two practical routes to the identification of tumour suppressor genes suggest themselves from the accumulated evidence above. The first would be to identify cellular components that interact with tumorigenic DNA viral gene products, and the second would be to clone genes that are deleted or mutated in inherited cancer susceptibility syndromes. It is very gratifying to find that both routes converge on the same gene products. This confirms the suspicion that there are a relatively small number of gene products with tumour suppressor functions.

The retinoblastoma gene (*Rb*)

The target gene (termed *Rb*), which is deleted in individuals with inherited retinoblastoma susceptibility, was isolated by positional cloning methods, taking advantage of the fact that many cases of retinoblastoma involved small regions of chromosomal deletion. It was rapidly confirmed that this gene was indeed responsible for all cases of inherited retinoblastoma examined and, crucially, the basic predictions of the two-hit hypothesis could be confirmed. Individuals at risk inherited a single mutated copy of the gene, and tumours derived from these individuals had inactivating mutations or deletions in both copies of the gene. Moreover, introduction of a wild-type *Rb* gene into tumour cells derived from retinoblastoma patients was able to restore normal growth. Finally, it was found that the *Rb* gene was also mutated in a variety of 'spontaneous' adult tumours, indicating that it might be involved in a variety of tumours aside from retinoblastoma.

The *Rb* gene encodes a nuclear-located phosphoprotein of molecular mass 110 kDa. The Rb gene product forms a complex with the DNA viral oncogene proteins SV40LT, E1A and HPVE6 in virally infected cells. The site of interaction with DNA viral oncoproteins is the 'pocket' domain, a region of the protein which has been found to be point mutated in some inactivating mutants of Rb. This implies that this region of the Rb protein may be of significance for its tumour suppressor functions, and the purpose of the association between the viral gene products and the pocket domain may be to inactivate or inhibit the association of Rb with normal cellular proteins. This focuses attention on the identity of cellular components that interact with Rb.

A number of Rb-interacting proteins have been identified by different routes. An important partner is a transcription factor, E2F, a multigene family of proteins which, in association with a heterodimerising partner factor DB, can activate expression of a number of genes involved in the mechanism of DNA synthesis, such as dihydrofolate reductase, the proliferating cell nuclear antigen (PCNA) subunit of DNA polymerase and, significantly, the cyclin E gene. This indicates that active E2F is involved in coordinating the onset of S phase by transcriptional activation of DNA synthetic machinery and activation of S-phase CDK activity. Indeed, consistent with this idea, forced expression of E2F in quiescent cells induces the premature induction of DNA synthesis.

What is the function of the E2F/Rb complex? E2F binds to Rb via the pocket domain and is, therefore, presumably released in the presence of DNA viral oncoproteins. This suggests that the E2F/Rb complex acts to suppress the transcription of E2F-dependent promoters and that, in order for S phase to occur, this suppression must be relieved. This indeed appears to be the case (Fig. 8.3). The interaction of Rb with E2F blocks the so-called activation domain of E2F, which is required for E2F-dependent activation of downstream promoters. E2F, in effect, tethers the Rb protein onto E2F-dependent promoters in the form of a transcriptionally silent complex. The suppression of transcription is further en-

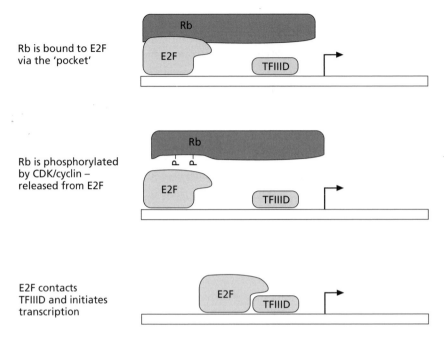

Rb is bound to E2F
via the 'pocket'

Rb is phosphorylated
by CDK/cyclin –
released from E2F

E2F contacts
TFIIID and initiates
transcription

Fig. 8.3 Mechanism of Rb action. Before activation of cyclin D dependent kinases, Rb is bound to E2F; amongst other things, this blocks access of E2F to the basal transcription machinery, such as the core factor TFIIID. Phosphorylation of Rb by cyclin D/CDK kinases causes release of Rb from the complex with E2F and consequent association of E2F with TFIIID. This leads to transcriptional activation of E2F-dependent genes involved in the onset of S phase.

hanced by the recruitment of histone deacetylase into the complex via association with Rb. This results in nucleosome assembly in the vicinity of E2F-dependent promoters and position-specific suppression of transcription.

The complex between Rb and E2F must therefore be dissociated or inactivated in some way for E2F activity to be manifest and S phase to proceed. This in turn indicates that Rb function must be controlled by some form of cell cycle-dependent regulatory mechanism.

Close examination of the protein in synchronous cell populations reveals that, just prior to the onset of S phase, it is phosphorylated at a number of sites, including the pocket domain. This phosphorylation is mediated by the cyclin D/CDK complex and results in activation of E2F activity, presumably by dissociation of the E2F/Rb repressor complex. In essence, Rb acts to block the programme of gene expression required for the onset of S phase by, *inter alia*, transcriptional repression of E2F-dependent promoters. This inhibitory activity can be relieved either by mutation and loss of Rb function, in association with DNA viral oncogenes, or, in the course of normal cell cycle progress, by phosphorylation by active CDK complexes (Fig. 8.4). In this context, Rb resembles a

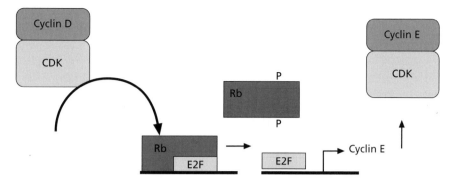

Fig. 8.4 Rb function is at the junction between two cyclin-dependent kinases (CDKs). Phosphorylation of Rb by the cyclin D/CDK complex leads to activation of E2F-dependent genes (see also Fig. 8.3). Among the genes is cyclin E, which leads to activation of the cyclin E/CDK complex; this orchestrates the onset of S phase.

'gate' at the end of the G1 phase and prevents progress through the cell cycle unless opened by the activity of CDK/cyclins. Blockage of Rb activity therefore removes a barrier to progress through the cell cycle.

Although this account provides a persuasive explanation for how *Rb* gene deletion may result in a propensity to form tumours, it is clear that there are features of E2F/Rb function that remain to be understood. Genetic ablation of E2F might be predicted to be fatal in view of its role in the entry into S phase. In fact, mice with this ablation are viable and fertile but exhibit an increased probability to develop tumours in later life. On these grounds, E2F would appear paradoxically to exhibit the properties expected of a tumour suppressor gene. This may reflect the action of other members of the E2F family in 'normal' cell cycles, or that the role of E2F is manifest only under specific cellular conditions. In addition, mice that are genetically deficient in Rb in fact expire during late gestation as a result of dramatic cell death in the nervous and haemopoietic systems. This indicates that, as with E2F, Rb function is seemingly not required in every cell cycle and Rb may have other functions in development and differentiation via association with non-E2F proteins.

p53

The second example of a tumour suppressor gene that we shall examine is *p53*, which was named on the basis of its discovery as a normal cellular protein of molecular mass 53 kDa, tightly associated with the DNA viral oncogene product SV40LT. It was subsequently found to interact also with the DNA viral oncogenes encoded by the human papilloma viruses. On these grounds, *p53* could be considered a candidate tumour suppressor gene. This was firmly established when it was discovered that the Li–Fraumeni syndrome resulted from inheritance of a *p53* gene mutation. In addition, *p53* gene mutation (by either dele-

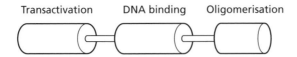

Fig. 8.5 The domain structure of p53. The N-terminal domain is involved in interaction with the transcription machinery; it also binds mouse double minute (MDM) and is the target for phosphorylation by kinases activated by DNA damage. The central domain is involved in sequence-specific DNA recognition and the C-terminal domain is required to oligomerise p53.

tions or point mutations) is an extremely frequent feature of many (but not all) naturally occurring tumours. Indeed, up to 85% of breast, lung and colon cancers were discovered to harbour *p53* gene mutations. Finally, mice that are genetically deficient in *p53* exhibit a greatly increased propensity to develop tumours, especially in response to genotoxic insults such as chemical carcinogens or ionising radiation. Taken together, this evidence clearly shows that normal p53 function acts to 'protect' against tumour formation but its activity is probably not required for every cell cycle.

What is the function of p53? Forced expression of wild-type p53 protein in cells leads to cell cycle arrest prior to S phase and eventual cell death. It is therefore a protein whose normal function is to arrest cell proliferation and kill cells.

How does p53 execute this function? Wild-type p53 is a nuclear-located protein found in the form of a tetramer. It can be considered to comprise three functional domains (Fig. 8.5). The C-terminal region is required for oligomerisation of the protein and binds preferentially to single-stranded DNA or DNA 'ends'. The central domain is involved in sequence-specific DNA recognition and its affinity for DNA is modulated by the phosphorylation of, or DNA binding to, the C-terminal domain. Finally, the N-terminal region of p53 interacts with other cellular proteins required for transcriptional activation and is another target for regulatory phosphorylation events. On this basis, p53 is a sequence-specific DNA-binding transcription factor whose activity is enhanced by association with single-stranded DNA breaks and other cellular partners. Indeed, many naturally occurring mutations of *p53* inhibit DNA binding by direct or indirect means and the SV40LT oncoprotein blocks the DNA-binding domain of p53. The cell cycle suppression functions of p53 must therefore result from its ability to regulate gene transcription.

How does *p53* act as a tumour suppressor gene? Perhaps not surprisingly in view of its biological functions, *p53* expression is very low in normal tissues due, amongst other things, to a very high turnover of the protein under normal conditions. In fact, in most cases, p53 protein is only detected in cells that are proliferating very rapidly (i.e. have a high proportion of the population in S phase) or in normal cells where a transient peak of expression is detected at the onset of S phase. However, p53 levels in the cell are dramatically elevated in response to

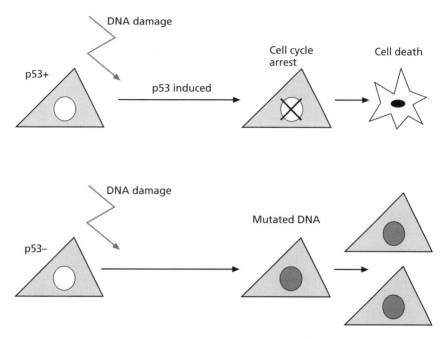

Fig. 8.6 p53 is the 'guardian of the genome'. DNA damage in cells with normal p53 leads to the activation of cell cycle arrest genes such as *p21* and *GAD45*. Prolonged expression of p53, such as occurs with failure to repair DNA damage, leads to the induction of apoptosis and death of the damaged cell. DNA damage in cells which lack p53 function leads to mutated DNA and the accumulation of mutations and chromosome abnormalities. These cells continue to proliferate.

treatments that result in DNA damage, such as exposure to ionising radiation, genotoxic carcinogens and, perhaps unfortunately, DNA-damaging cancer chemotherapeutic agents like adryomycin. This finding provides a vital clue. It suggests that *p53* acts as 'the guardian of the genome'—in other words, *p53* is a gene which is induced in response to DNA damage and acts to block cell cycle progression (Fig. 8.6). Under these circumstances, cells with genomic damage cannot divide and the action of p53 suppresses the accumulation of chromosomal rearrangements and further mutations. If the DNA damage is repaired, p53 levels fall and progress through the cell cycle proceeds. If the DNA damage is not repaired, p53 eventually acts to kill the cell. In other words, p53 is a central mechanism that monitors the integrity of the genome in each cell cycle and acts to prevent cells harbouring genome rearrangements from further proliferation. It follows from this that a loss of p53 function should lead to genome instability and the accumulation of mutations. It should be recalled that this is both one of the hallmarks of established cell lines and the underlying basis of many types of proto-oncogene activation.

p53 is therefore a transcription factor that has the property of halting cycle progress when expressed at appropriate levels in cells with damaged DNA. It fol-

lows that this inhibition of cell cycle progress (and cell death) may well be mediated by the gene targets for p53 transactivation. Considerable effort has been
expended on identifying genes whose transcription is regulated by p53 (Fig.
8.7). One target for p53 induction is a protein called GADD45. GADD45 binds to
PCNA (which is induced by E2F), a regulatory subunit of the DNA polymerase

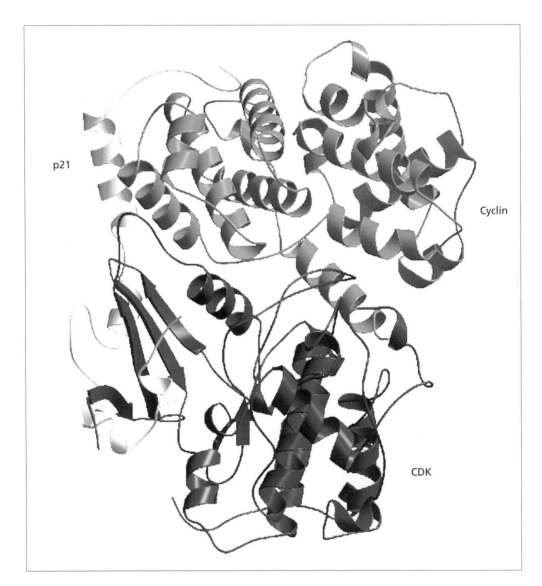

Fig. 8.7 The three-dimensional structure of the complex between a peptide derived from p21
and the cyclin/cyclin-dependent kinase (CDK) complex (see also Fig. 6.5). The peptide binds
both the cyclin and the kinase and, as a result, the CDK activation loop is held in a 'closed'
position, blocking kinase activity.

complex. The association of GADD45 with PCNA has the interesting property of blocking the activity of DNA polymerase in DNA replication but not DNA repair. In theory, the effect of GADD45 expression should therefore be to halt DNA replication but not repair.

Amongst the most significant p53-inducible gene products is p21 (also known as Waf 1), which appears to be the agent responsible for most of the effects of p53 expression on cell cycle progress. Forced expression of p21 in a proliferating cell induces a halt to cell cycle progress that is very similar to that induced by p53. p21 is a CDK inhibitor that blocks cyclin/CDK activity by binding to the cyclin/CDK complex in a manner which blocks the activation loop of the CDK (Fig. 8.7). The immediate result of p21 action is to block the phosphorylation of the E2F/Rb complex, in effect closing the gate to entry into S phase, and to block the cyclinE/CDK complex, thereby blocking the orchestration of S phase functions.

The biological effects of p53 are mediated by association with other cellular proteins that associate with the N-terminal activation domain. As mentioned above, p53 associates with the SV40LT antigen in virally infected cells, but p53 has also been found to bind two other DNA viral oncoproteins. Adenovirus E1A binds the N-terminal transactivation domain of p53, thereby inactivating the transcriptional activation functions of p53, and the E6 gene of human papillomavirus can also bind p53, promoting its degradation. By analogy with the association of viral oncoproteins with Rb described above, it may be presumed that the function of the viral proteins is generally to inactivate p53 activity; in this case, presumably to promote the process of viral DNA replication during the infectious cycle. Pursuing this analogy further, it might also be that viral proteins compete in binding with normal cellular proteins. In *in vitro* studies a number of proteins have been found to bind p53. These include RNA helicases and elements of the basal transcription apparatus.

However, a potentially significant p53 partner is the MDM2 (for mouse double minute) protein (Fig. 8.8). This was originally identified in murine sarcomas as a putative oncogene activated by gene amplification. MDM2 protein binds to the N-terminal transactivation domain of p53 and blocks its biological functions. The activation of MDM2 by gene amplification suggests that it exerts its actions by titrating out biologically active p53 protein, thereby emulating p53 inactivating mutations. In addition, MDM2 is also a target for *p53* gene induction, suggesting the existence of some form of inhibitory feedback loop whereby p53 induction of MDM2 expression reverses the inhibitory control exerted by p53. MDM2 may also be implicated in modulating the functions of p53 under normal physiological conditions, since MDM2-deficient mice die soon after implantation in a p53-dependent fashion. This suggests that a function of MDM2 is to neutralise a potentially toxic effect of p53 on embryonic development.

The overall model for p53 action (Fig. 8.8) therefore suggests that the p53 transactivation function is activated upon exposure to single-stranded DNA or DNA 'ends'. This situation can occur either during the firing of replicons in nor-

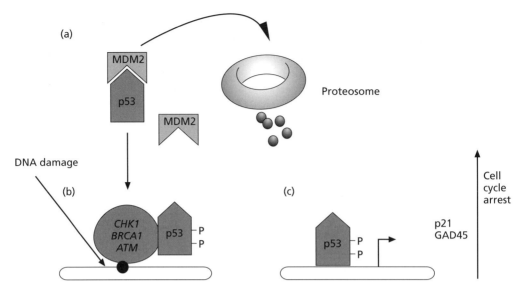

Fig. 8.8 The mechanism of p53 activation by DNA damage. (a) Under normal circumstances, p53 is bound via the N-terminal domain to MDM2. This induces ubiquitinylation of p53 and proteolytic degradation via the proteosome machinery. (b) DNA damage results in the activation of protein kinases and includes the tumour suppressor genes *CHK1*, *ATM* and *BRCA1*. (c) This leads to the phosphorylation of p53 in the N-terminal domain, release from MDM2, stabilisation of the p53 protein and activation of target gene transcription.

mal scheduled DNA synthesis or as a result of unscheduled DNA damage. Activation of p53 occurs via two mechanisms: the first is protein stabilisation due to relief of MDM-mediated p53 destruction. Second, it is also clear that p53 function requires phosphorylation, and emerging evidence indicates that this occurs via activation of protein kinases that are recruited to regions of damaged DNA. It might be predicted that these kinases would also have tumour suppressor gene function, since an inability to phosphorylate p53 would result in loss of p53 function. Indeed, a number of tumour suppressor genes, such as *BRCA1*, *CHK1* and *ATM*, are involved in the pathway leading to p53 phosphorylation. Finally, the two mechanisms are connected in that phosphorylation of the N-terminal region of p53 blocks the association with MDM.

As a consequence of p53 activation, a number of downstream target genes, including *p21* and *GADD45*, are transcriptionally activated. The collective activity of these genes accounts for the arrest to progress through the cell cycle. When DNA repair is completed, p53 degradation increases, p53 protein levels drop and progress through the cell cycle resumes. p53 can therefore be regarded as an example of a 'checkpoint control' protein (see Chapter 6) whose primary function is to suppress cell cycle control at critical phases of the cell cycle and prevent the propagation of cells with damaged DNA. It can therefore be seen that p53, whilst not required for cell progress, is intimately involved with cell cycle machinery. A long-term consequence of the inhibition of p53 function is the accumulation of genomic alterations, including mutations and rearrangements that lead to oncogene activation. The means by which p53 achieves cell cycle arrest is to exploit, via p21, the cyclin/CDK core cell cycle machinery and, indirectly, the actions of another tumour suppressor gene, *Rb*.

CDK inhibitors

The p53-inducible CDK inhibitor is one member of a larger family of proteins with CDK inhibition functions. All CDK inhibitors cause G1 arrest when overexpressed in transfected cells. This is, in effect, a forced version of premature senescence. Moreover, loss of CDK inhibitor function also prevents cells from entering senescence and renders them susceptible to single-hit transformation by activated oncogenes. Loss of CDK inhibitor function by mutation or inhibition by DNA viral oncogenes is therefore the major biochemical event underlying the phenomenon of establishment described in Chapter 1. Induction of CDK inhibitor function also provides the basis for the reversible growth arrest induced by the TGFB family. It should therefore be predicted that this class of proteins would exhibit tumour suppressor function. Indeed, direct or indirect blockage of CDK inhibitor pathways proves to be a central theme in the activity of tumour suppressor genes.

p21 is the prototype member of the Kip family of CDK inhibitors whose other members are p27 and p57. The difference between these inhibitors resides in the specificity of the CDK family members that are inhibited with p57 and p27 exhibiting broad-spectrum CDK inhibition, whereas p21 inhibition is restricted to the G1 phase CDK4/cyclin D complex. There is very good evidence that p27 activity plays a normal role in cell cycle control. p27 protein levels are elevated in quiescent cells but fall when cells are stimulated to progress through the cell cycle. Mice that are genetically deficient in p27 are larger than their normal counterparts as the result of an increased number of cells in every organ, indicating that p27 activity is involved in the regulation of tissue mass in normal development. p27 displays all the hallmarks of a classical tumour suppressor gene, since p27-deficient mice also are susceptible to the production of pituitary tumours.

The second family of CDK inhibitors are the ink4 family, p16, p15, p18 and p19; these are specific inhibitors of the G1 phase cyclin-dependent kinases CDK4 and CDK6, which play essential functions in the regulation of Rb function. The *p16 ink4A* is a frequently mutated gene in many types of human tumours and must rank in status with p53 as a major target for mutation in human cancer. Unusually, the *p16* gene can be translated in two alternative reading frames to produce two different gene products: the CDK inhibitor protein p16 and a second protein p19ARF, an alternative reading frame protein that also blocks cell proliferation. p19ARF acts directly on p53 function via inhibition of the action of MDM2 and concomitant activation of p53 growth arrest functions.

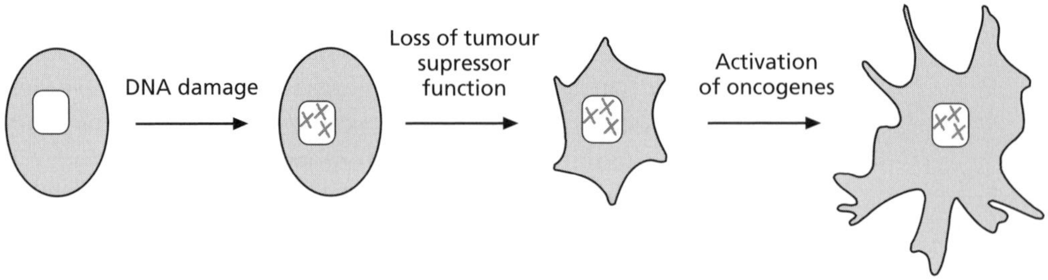

Fig. 8.9 A generalised model of tumour formation. Mutation of tumour suppressor genes leads to the inactivation of DNA damage-sensing mechanisms and loss of cell cycle checkpoint control. This leads to genetic instability and a loss of the ability to enter senescence. The mutation of proto-oncogenes resulting from DNA damage leads to oncogenic activation, induction of cell cycle progress and an increase in tumour cell mass.

The *p16* gene is therefore plugged into both the Rb and p53 tumour suppressor pathways.

Tumour suppressor genes in perspective

The genetic analysis of malignancy has revealed the existence of additional pathways which regulate cell cycle progression. These take two forms. One is a block to entry into S phase, as exemplified by the activity of the Rb protein. This block must be relieved for progress into S phase and the remainder of the cell cycle to occur. The second is the existence of checkpoint controls which, whilst not essential for cell cycle progression, have the ability to block progress in the event of DNA damage.

Both pathways centre around suppression of the CDK pathways of the cell cycle engine. The majority of malignancies involve direct or indirect inactivation of the biochemical pathways that inhibit the activity of CDKs. The function of activated oncogenes in promoting indefinite cell proliferation can only be revealed when these pathways have been inactivated. This provides a clear mechanistic basis for understanding the requirement for multiple mutations to occur on the route to malignancy.

The analysis of tumour suppressor gene function has also revealed an intimate connection with the processes leading to activation of proto-oncogenes by mutation or genome rearrangement. This is due to the fact that the cycle engine is also controlled by checkpoint control mechanisms, as exemplified by the p53 protein. Checkpoint control mechanisms block progress though the cell cycle when DNA damage occurs via the CDK inhibitor pathway and direct intervention in the mechanism of DNA replication. Removal of checkpoint controls allows the destabilisation of genetic integrity and promotes the propagation of mutations and genome rearrangements; this inevitably leads to proto-oncogene activation and promotion of progressive cell growth in the form of a tumour.

This leads to a unifying view of the genetic targets that must be modified in the process of tumour formation (Fig. 8.9). The first are components of the pre-restriction point signalling machinery. The second are components involved in the sensing and repair of damaged DNA; and the third control the activities of the CDKs which orchestrate the main events of the cell cycle engine.

Chapter 9: Cell Survival

Introduction

The preceding chapters have been devoted to analysis of the molecular mechanisms governing cell replication — the process by which two cells are produced, where one was present before. This process is a fundamental, if not defining, feature of all living organisms. For multicellular organisms the size of cell populations is integrated and tightly controlled, and it is an amazing fact that the relative cell numbers in tissues and organs varies little between individuals within the same species. This is brought about by the coordinated control of cell proliferation, which has been discussed in previous chapters, and the coordinated death of cells, which is the subject of this chapter. Regulated cell death is a defining feature of multicellular organisms. It reflects a unique quality of multicellularity which is that the fate of individual cells is subsumed to the overall requirements of the organism.

Cell death is a common feature in the development of many tissues. For example, from all the cells generated during the development of the nervous system, over 50% are fated to die without contributing to the final structure. Moreover, during spermatogenesis in man, many spermatagonia die before giving rise to functional sperm. In the female germ line about 30 000 oogonia are formed during embryonic life but only about 3000 survive to be able to mature in response to hormones. Of these, perhaps only 10% survive to produce fertilisable eggs during the lifetime of an individual. Perhaps even more staggering is the elimination of more than 95% of lymphocytes by cell death during their development. The scale and ubiquity of cell death in the biology of multicellular organisms suggests that something special (and conserved across species) must be happening.

Cell death is important not only in normal physiology but also in pathology. Thus, tumours can, in principle, increase in mass due to cell proliferation but they could also shrink in mass by cell death. Is it possible that the abnormal failure of a cell to die could also contribute to the growth of tumours? This leads to the idea that cell death may not be the result of the failure or disintegration of molecular processes but, rather, the consequence of the controlled activity of a cellular suicide machine.

The appearance of apoptosis

Cell death can be found in every multicellular organism and a remarkable property is that in every case, whatever the cell type or the species, it exhibits the

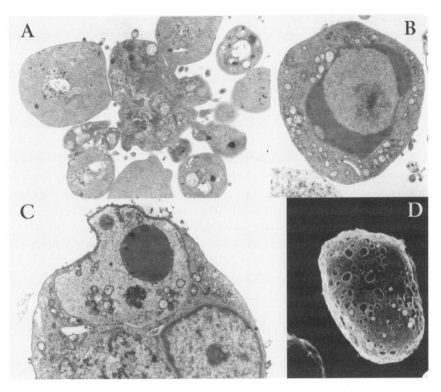

Fig. 9.1 The morphological hallmarks of apoptosis. (A) Transmission electron micrograph of an apoptotic fibroblast with fragmentation of the cell into apoptotic bodies. (B) The early stages of apoptosis, showing chromatin condensation. (C) A phagocytosed apoptotic cell. (D) Scanning electron micrograph of an apoptotic cell showing loss of microvilli and gaping pits formed by fusion of the endoplasmic reticulum with the plasma membrane. (Courtesy of Professor A. Wylie; reprinted with permission from Rich, T., Watson, C.J. & Wylie, A. (1999) *Nature Cell Biology* **1**, 69–73. Copyright (1999) Macmillan Magazines Limited.)

same morphological features (Fig. 9.1). These include a characteristic blebbing or 'boiling' of the plasma membrane, condensation of chromatin into a single 'lump' inside the nucleus, fragmentation of DNA leading to characteristic laddering, and the induction of 'megapores' in mitochondria which releases their contents into the cytoplasm. Following the process of cell death in real time reveals that it is remarkably rapid, occupying less than an hour from the first appearance of membrane blebbing to complete disintegration of the cell. In normal tissues, dead cells are rapidly removed by macrophages, leaving no trace of their existence. These stereotype features of the process have led to the general term 'apoptosis' to denote the controlled destruction of a cell by an active biochemical mechanism.

Why should apoptosis result from activation of a specific biochemical machine? The two central reasons for supposing that apoptosis results from activation of a biochemical process are, firstly, that it can be regulated by external stimuli and, secondly, that mutations can be recovered in specific genes which regulate the process. Neither of these would be predicted from a mechanism which simply involved dysfunction or random failure of cellular functions.

It has been noticed for over a hundred years that one of the most obvious consequences of placing cells from a normal tissue into culture is the rapid occurrence of massive cell death in the majority of cells. This was, for a long time, believed to reflect either an unfortunate failure to provide the correct nutrient conditions or the presence of noxious compounds in the culture medium. However, this kind of cell death could be ameliorated in the short term by inhibition of protein synthesis or, in the longer term, by the presence of specific growth factors in the culture environment. Indeed, for many cell types, specific growth factors are continuously required to be present in order for cells to survive. This suggests that growth factors not only signal cells to proliferate but might also be involved in providing a second type of signal to cells which, in effect, says 'stay alive'. Removal of this signal results in immediate cell suicide. Thus, for example, certain types of neuronal cells are utterly dependent upon the presence of specific neurotrophic growth factors, such as nerve growth factor, or ciliary neurotrophic factor for survival in culture. This leads to the idea that the size of cell populations *in vivo* might in fact be controlled by the availability of specific growth factors whose function is to keep cells alive by inhibiting apoptosis. If these factors are present in limiting amounts or in specific locations, cells located in the 'wrong' place, or in inappropriate amounts, would simply die. This is, in fact, the case. For example, the extensive death of neuronal populations that is observed to occur in normal development can be prevented by adding excess neurotrophic factors, indicating that the size of neuronal populations is controlled by limiting access to specific apoptosis-inhibiting factors.

The idea that cell death is controlled by extracellular signals can be turned on its head. Perhaps there are receptor-dependent factors which perform the opposite functions, namely to signal cells to die. This type of mechanism is particularly prevalent in the immune system where specific extracellular ligands, such as tumour necrosis factor (TNF) and its relatives, act to eliminate cells during normal lymphopoiesis. A consequence of blocking the action of these specific death-inducing factors is pathological proliferation of lymphoid cell populations. This implies the existence of receptor-mediated death pathways within the cell.

As described above, apoptosis is a reproducible feature of the normal development of multicellular organisms. A particularly informative example of developmental cell death occurs in the nematode worm *C. elegans*. The advantage of the worm for this study is that it is composed of only a few hundred cells in total and develops in a highly reproducible manner. The worm also has very tractable genetics. It was noticed during the normal development of the worm that specific cells underwent apoptosis in a very reproducible manner (Fig. 9.2) and that, in some cases, it was possible to recover mutants (so-called ced mutations) in which cell death did not occur. This finding is of considerable interest. It indicates that the process of cell death is regulated by specific gene products

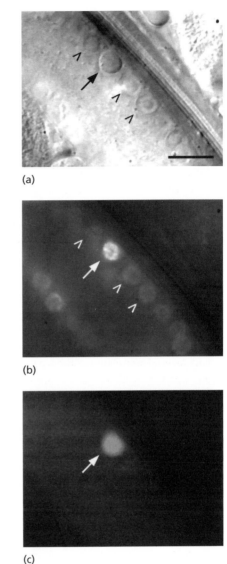

(a)

(b)

(c)

Fig. 9.2 Cell death in the development of *Caenorhabditis elegans*. An apoptotic cell (arrow) in the *C. elegans* germ line can be distinguished from surviving neighbours (arrowheads) by: (a) its altered shape (detected by differential interference contrast (DIC) light microscopy); (b) its condensed nuclear chromatin (stained with the DNA stain Hoechst 33342); and (c) selective uptake of the vital dye acridine orange. (Courtesy of Michael Hengartner, Cold Spring Harbor Laboratory.)

and is inhibited if these products are missing. This adds to the evidence that apoptosis is an active process and paves the way to the isolation and genetic analysis of genes which participate in the mechanism of apoptosis.

Molecular mechanisms of apoptosis

The evidence described above clearly points to the existence of a regulable molecular process controlling apoptosis. In many ways this pathway can be considered analogous to the pathways involved in cell proliferation, since it involves activation of intracellular pathways by ligand-mediated receptor activation.

Pursuing the analogy with the cell cycle one step further, it is convenient to divide cell death into two main processes: (i) the 'death engine' or the core processes responsible for the characteristic features of apoptosis; and (ii) the 'trigger' or signalling pathways that activate the death engine.

The death engine

As mentioned above, the existence of specific mutations which block apoptosis proves to be a very valuable starting point for identifying components of the apoptosis machinery. One of the first mutations in *C. elegans* to inhibit cell death was a gene called *ced-3*. This gene appeared to encode a gene product that lay at the heart of apoptosis, since *ced-3* mutants were unable to activate apoptosis in all cells fated to die in normal development.

Cloning and characterisation of the *ced-3* gene product revealed that it encoded a protease of a very specific family, now called the caspases (by virtue of the presence of cysteine in the active site and an affinity for substrates containing an aspartic acid residue). The caspases prove to be a large family of highly conserved cytoplasmic proteases that represent the most prominent mechanism for destruction of the cell.

The caspase family of proteases share similarities in amino acid sequence, structure and substrate specificity (Fig. 9.3). They are present in the cell in the

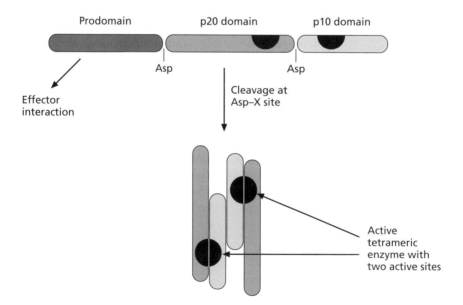

Fig. 9.3 Activation of caspases. The classical caspase contains three domains: the C-terminal prodomain, which interacts with effectors, and the p10 and p20 domains, which contain the active sites. Cleavage at the aspartate (Asp) residues flanking the p20 domain releases the prodomain and leads to the formation of an active tetrameric caspase enzyme formed from two p20 and two p10 subunits.

form of proteolytically inactive proenzymes and the conversion of the proenzyme to the catalytically active protease represents a crucial step in the initiation of apoptosis. Caspases comprise three domains: an N-terminal domain, which represents the major region of variation between different caspase family members, and the conserved catalytic core, which is composed of two protein domains—the large and small subunits. Studies of substrate preferences show that different caspase family members exhibit different protein substrate specificities, indicating that each enzyme is potentially involved in the proteolysis of different cellular targets.

Since caspases are present continuously in cells in an inactive state but can be mobilised rapidly upon the receipt of a pro-apoptotic signal, it is clear that their mechanism of activation is a vital feature of the apoptotic process. Caspase activity can be unleashed via two basic mechanisms: the first is proteolytic cleavage operating in a cascade manner. Thus, an 'initiator' caspase cleaves an effector caspase; this activates its proteolytic activity against specific substrates, which include other cascade family members. The kinetics and specificity of this proteolytic cascade are controlled by the differing substrate specificities of caspase family members.

This mechanism does, however, pose the problem of how the initiator caspase is activated. It is interesting to note that a second mechanism for activating caspases is oligomerisation. Caspase proenzyme activity can be activated by fusion to dimerisation domains of other protein partners. This suggests that caspases are latent in the cell because they exist at low concentrations as monomers. The activation mechanism therefore serves to bring two or more caspase precursors in close proximity, allowing for intermolecular autoproteolytic activation.

It is clear that other means of inhibiting caspase activity exist. There are, for example, naturally occurring caspase inhibitors, some of which are 'decoys' or catalytically inert versions of caspases. In addition, certain viruses encode proteins that block caspase function, presumably to facilitate the process of viral infection. It is also possible, as we shall see, that caspases are sequestered or compartmentalised into specific cellular compartments and released upon receipt of a pro-apoptotic signal.

If caspases are required for apoptosis to occur, it follows that many of the typical manifestations of the apoptotic process must result from caspase-induced proteolysis. One way of looking at this is that caspases contribute to apoptosis through direct disassembly of cell structures, such as the destruction of the nuclear lamina (a rigid structure that underlies the nuclear membrane and is involved in chromatin organisation). During apoptosis, lamins are cleaved at a single site by caspases, causing the nuclear lamina to collapse and contributing to chromatin condensation. Another important substrate is gelsolin, which severs actin filaments in a regulated manner. Caspase cleavage generates a fragment of gelsolin that is constitutively active, thereby leading to collapse of the actin cytoskeleton. The characteristic cleavage of DNA in apoptotic cells is brought about by activation of an apoptosis-specific DNAse through cleavage of a specific inhibitor protein. In general terms, a survey of caspase substrates indicates that the process of caspase-mediated cell destruction resembles a tightly controlled military operation. Contacts with surrounding cells are removed, the cytoskeleton is collapsed, DNA replication and repair are blocked, the nuclear structure is collapsed and the DNA itself is efficiently degraded. The use of specific caspase inhibitors permits, to a certain extent, the identification of the involvement of different caspase family members in different parts of the destruction process. It also emerges from these studies that at least some diagnostic features of apoptosis, in particular the characteristic blebbing or boiling of the plasma membrane, which is the first indication of cell death, are unaffected by inhibition of caspase activity, suggesting that other mechanisms may be involved.

The death trigger

If a caspase-mediated proteolytic cascade represents the main (but not sole) engine of cell destruction, what pulls the trigger? The evidence at this point is fragmentary in nature and we await a fully integrated account of the mechanisms involved. What is clear, however, is that several different pathways can contribute, by apparently different routes, to the activation of the death engine.

One informative route has been the molecular dissection of the action of 'death inducing' cytokines of the TNF family. The functions of TNF family members are executed by association with a specific family of cell surface receptors (Fig. 9.4). These act in a slightly different fashion to growth factor receptors in that the ligands are essentially trimeric in nature, and so association of a TNF family ligand with a specific receptor induces the formation of a trimeric receptor complex. Much attention has been devoted to mutagenesis of the cytoplasmic domains of TNF family receptors with the objective of identifying regions required for activation of the death programme. This has led to the identification of the 'death domain', a region of the receptor which interacts with specific protein adaptors (Fig. 9.4). A key TNF family adaptor protein is FADD, which associates with oligomerised receptor cytoplasmic domains, either directly or indirectly, via another adaptor, TRADD. FADD binds a specific member of the caspase family, caspase-8. A consequence of ligand-mediated receptor trimerisation is FADD-mediated oligomerisation of caspase-8, thereby leading directly to proteolytic activation. The TNF family of receptors is therefore coupled directly to the death engine, which is activated by receptor trimerisation.

The Bcl-2 family

The introduction posed the idea that inhibition of cell death could be a viable mechanism of oncogenesis. If this were true, one might expect to identify

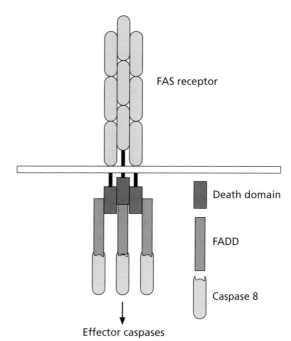

FAS receptor

Fig. 9.4 Activation of caspases by the tumour necrosis factor (TNF) family ligand, FAS. Interaction of the FAS ligand induces trimerisation of the FAS receptor. Caspase 8 is associated with the FAS receptor cytoplasmic domain via the adaptor protein, FADD. Oligomerisation of the caspase via FAS receptor trimerisation induces autocatalytic cleavage (see also Fig. 9.3) and caspase activation.

Death domain

FADD

Caspase 8

Effector caspases

'oncogenes' which act to block cell death rather than promote cell proliferation. This proves to be the case. The oncogene *Bcl-2* (for B-cell leukaemia 2) was isolated as a gene activated by chromosome translocation in human lymphoma. Analysis of *Bcl-2* action unexpectedly revealed that it was able to permit the survival of cytokine-dependent haematopoietic cells, in a quiescent state, in the absence of cytokine. The activity of the Bcl-2 gene product therefore appeared to be engaged in the machinery of cell survival rather than cell proliferation. Moreover, this finding indicates that Bcl-2 activity lies downstream of, or in some way emulates, the ability of cytokine-mediated signalling to promote cell survival.

The mechanism of Bcl-2 induced oncogenesis appears to involve activation or elevation of gene expression, suggesting that the normal function of Bcl-2 is to promote cell survival or inhibit apoptosis. The biological significance of Bcl-2 took on an extra dimension, however, when it was discovered that the product of *C. elegans* cell death mutant *ced-9* was in fact the worm homologue of the human *Bcl-2* gene. Further analysis reveals that *Bcl-2* is the prototype member of a much larger, structurally related gene family that has biological effects on cell survival.

Studies of cells transfected with diverse Bcl-2 family members indicate that, broadly speaking, they fall into two functional classes with respect to their effects on cell survival. One class, exemplified by the *Bcl-2* prototype and including other members (such as *Bcl-xl*, *Bcl-w* and the *ced-9* homologue) are 'survivors'. That is to say, they function in, for example, transfected cells to promote cell survival. The other class, exemplified by genes such as *Bax*, *Bak* and

Bok, can be considered to be 'killers'. In other words, they function to promote apoptosis. The Bcl-2 family of proteins form heterodimeric associations with other family members, including pairings between 'survivors' and 'killers'. This suggests that the effects of the Bcl-2 family on cell survival might depend on the relative ratio of different family members, the biochemical specificity of dimer partners and the resulting preferential composition of Bcl-2 family protein complexes.

Bcl-2 is located in the cell in association with internal membranes, especially the outer membrane of the mitochondrion. In addition, the three-dimensional structure of Bcl-2 is intriguingly reminiscent of the pore-forming complexes of certain bacterial toxins. This suggests that the Bcl-2 family might form regulable 'megapores' in the mitochondrial (and other) membranes which, when opened, would lead to the efflux of mitochondrial contents and release of outer membrane proteins, such as cytochrome c; this is a characteristic feature of cells undergoing apoptosis. In this model the 'leakiness' of cell membranes would be controlled either by the composition of the pore-forming complex or by direct regulation of the Bcl-2 family structure. The evidence for this model is currently circumstantial. It is clear that Bcl-2 family proteins have pore-forming abilities in artificial membranes and there are indications that these channels exhibit some ion selectivity. It is also clear that Bcl-2 family members are potentially subject to regulation by external signals generated by growth factor-mediated signalling, since at least some family members, such as the killer protein Bad, become phosphorylated as a result of activation of the PI3K/AKT pathway. In the case of Bad, phosphorylation appears to inhibit its association with Bcl-xl, which might be expected to have pro-apoptotic consequences.

A seemingly independent mode of function of the Bcl-2 family is, however, suggested from further studies of death-deficient mutants of *C. elegans*. Genetic studies have shown that a third gene, *ced-4*, is required for the survival functions of the Bcl-2 homologue, *ced-9*. In the absence of ced-9, the product of the ced-4 gene has the ability to directly activate the caspase machinery in the form of ced-3. This is due to the formation of a complex between ced-4 and ced-3 and concomitant caspase activation by oligomerisation. A mammalian homologue of the ced-4 protein, Apaf-1, has been identified, and leads to a model which directly links the Bcl-2 family into the caspase-dependent apoptosis machine (Fig. 9.5). This supposes that pro-survival members bind Apaf-1 and, perhaps by physical sequestration, prevent it from activating procaspases. A death signal could provoke interaction of a 'killer' Bcl-2 family member which inhibits the neutralisation of Apaf-1. As a consequence, Apaf-1 can then bind to procaspase-9, promoting dimerization and protease activation by autocatalysis.

The 'dual key' hypothesis: *myc* oncogene and IGFs

It may be recalled from Chapter 7 that the precise biological functions of the *myc*

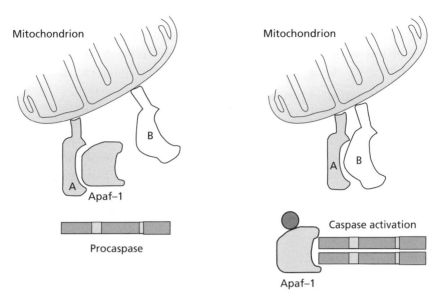

Fig. 9.5 Model for Apaf-1 regulation by the Bcl-2 family. A pro-survival Bcl family member (A) binds Apaf-1 and prevents it from activating the procaspase-9. A death signal provokes interaction of a pro-apoptotic family member (B), preventing it from neutralising Apaf-1. The formation of the A/B complex induces mitochondrial permeability and the release of cytochrome c. In the presence of cytochrome c released from mitochondria and ATP, Apaf-1 can then bind to procaspase-9 and promote its dimerisation and activation by autocatalysis. Caspase-9 subsequently activates effector caspases. (Adapted with permission from Adams, J. & Corey, S. (1998) *Science* **281**, 1322–1326. Copyright (1998) American Association for the Advancement of Science.)

oncogene are enigmatic, despite clear involvement in oncogenesis via its transcriptional activation functions. Indeed, expression of *myc* in primary cells, in the presence of growth factors, has immortalising activity, indicating that in this situation *myc* is engaging (presumably indirectly) with the cyclin/CDK cell cycle engine. However, a surprising feature of the *myc* gene is that it has also been implicated in the induction of apoptosis. At first sight this might seem to be an unusual manifestation of oncogenic activity. If the *myc* gene is expressed in quiescent cells, at admittedly supra-physiological levels, it induces apoptosis via a 'classical' caspase-dependent mechanism that requires the transactivation functions of the myc protein. This death-inducing activity of myc can be suppressed in the presence of specific growth factors, especially those of the IGF family (Chapter 2), which activate the PI3K/AKT pathway and impinge on the Bcl-2 mechanism. This dual activity of myc has led to the idea that apoptosis and cell cycle progression might in fact be linked. The 'dual key' hypothesis suggests that induction of progress through the cell cycle primes the apoptotic process and that apoptosis is executed, via the *myc* gene product, unless cell survival signals, mediated by molecules such as Bcl-2, are concurrently present. A logical outcome of this mechanism is the idea that, in normal tissues, a cell would be unable to proliferate without survival signals from neighbouring cells, since the activation of cell proliferation would inevitably result in cell death.

Apoptosis in perspective

Cell death is a ubiquitous and physiologically important determinant of the size of cell populations in multicellular organisms. It requires activation of a 'suicide machine' which involves a proteolytic cascade that systematically dismantles the cell architecture. This cascade is tightly regulated by external signals which can induce either cell death or cell survival. The survival signals seem to function, at least in part, via cellular sensors of the Bcl-2 family of proteins. Finally, the ability of classical oncogenes, such as *myc*, and tumour suppressor genes, such as *p53*, to impinge on the survival pathway indicates that the ability to induce cell death is an intrinsic feature of the mechanisms that control cell multiplication.

Further Reading

The literature relating to the subject of this book is vast. Here are some recommendations for further reading. This is not a scholarly compilation but simply papers I have enjoyed reading.

Biology of cell proliferation

Edgar, B.A. (1999) From small flies come big discoveries about size control. *Nature Cell Biology* **1**, E191–E193.

Holley, R.W. (1975) Control of growth of mammalian cells in cell culture. *Nature* **258**, 487–490.

Holley, R.W. & Kiernan, J.A. (1968) Contact inhibition of cell division in 3T3 cells. *Proceedings of the National Academy of Sciences USA* **60**, 300–304.

Pardee, A.B. (1974) A restriction point for control of normal animal cell proliferation. *Proceedings of the National Academy of Sciences USA* **71**, 1286–1290.

Pardee, A.B., Dubrow, R., Hamlin, J.L. & Kletzien, R.F. (1978) Animal cell cycle. *Annual Review of Biochemistry* **47**, 715–750.

Rossow, P.W., Riddle, V.G. & Pardee, A.B. (1979) Synthesis of labile, serum-dependent protein in early G1 controls animal cell growth. *Proceedings of the National Academy of Sciences USA* **76**, 4446–4450.

Signalling mechanisms

Darnel, J. (1998) STATS and gene regulation. *Science* **277**, 1630–1635.

Fambrough, D., McClure, K., Kazlauskas, A. & Lander, E.S. (1999) Diverse signalling pathways activated by growth factor receptors induce broadly overlapping, rather than independent, sets of genes. *Cell* **97**, 727–741.

Heldin, C.H. (1995) Dimerization of cell surface receptors in signal transduction. *Cell* **80**, 213–223.

Hemmings, B.A. (1998) Akt signalling—linking membrane events to life and death decisions. *Science* **275**, 628–630.

Hunter, T. (2000) Signalling—2000 and beyond. *Cell* **100**, 113–127.

Lemmon, M.A., Ferguson, K.M. & Schlessinger, J. (1996) PH domains: diverse sequences with a common fold recruit signalling molecules to the cell surface. *Cell* **85**, 621–624.

Massague, J. (1998) TGF-beta signal transduction. *Annual Review of Biochemistry* **67**, 753–791.

Mohammadi, M., Schlessinger J. & Hubbard, S.R. (1996) Structure of the FGF receptor tyrosine kinase domain reveals a novel auto-inhibitory mechanism. *Cell* **86**, 577–587.

Pawson, T. & Nash, P. (2000) Protein–protein interactions define specificity in signal transduction. *Genes and Development* **14**, 1027–1047.

Pawson, T. & Scott, J.D. (1997) Signalling through scaffold, anchoring, and adaptor proteins. *Science* **278**, 2075–2080.

Price, M.A., Hill, C. & Treisman, R. (1996) Integration of growth factor signals at the c-fos serum response element. *Philosophical Transactions of the Royal Society of London, Series B* **351**, 551.

Rameh, L.E. & Cantley, L.C. (1999)The role of phosphoinositide-3-kinase lipid products in cell function. *Journal of Biological Chemistry* **274**, 8347–8350.

Toker, A. & Cantley, L.C. (1997) Signalling through the lipid products of phosphoinositide-3-OH kinase. *Nature* **387**, 673–677.

Vishwanath, R., Iyer, Eisen, M.B *et al.* (1999) The transcriptional program in the response of human fibroblasts to serum. *Science* **283**, 83.

Weiss, A. & Schlessinger, J. (1998) Switching signals on or off by receptor dimerization. *Cell* **94**, 277–280.

Wrana, J. (2000) Regulation of Smad activity. *Cell* **100**, 189–192.

Cell cycle checkpoints and the cell cycle engine

Elledge, S.J. (1996) Cell cycle checkpoints: preventing an identity crisis. *Science* **274**,1664–1672.

Lee, M.G. & Nurse, P. (1987) Complementation used to clone a human homologue of the fission yeast cell cycle control gene cdc2. *Nature* **327**, 31–35.

Naysmith, K. (1996) Putting the cell cycle in order. *Science* **274**, 1643–1645.

Pines, J. (1999) Four-dimensional control of the cell cycle. *Nature Cell Biology* **1**, E73–E79.

Roberts, J.M. (1999) Evolving ideas about cyclins. *Cell* **98**, 129–132.

Sherr, C.J. (1996) Cancer cell cycles. *Science* **274**, 1672–1677.

Sherr, C.J. & Roberts, J.M. (1999) CDK inhibitors: positive and negative regulators of G1-phase progression. *Genes and Development* **13**, 1501.

Oncogenes, tumour suppressor genes and cancer

Agarwa, M.L., Taylor, W.R., Chernov, M.V., Chernova, O.B. & Stark, G.R. (1998) The p53 network. *Journal of Biological Chemistry* **273**,1–4.

Dynlacht, B.D. (1997) Regulation of transcription by proteins that control the cell cycle. *Nature* **389**, 149–152.

Dyson, N. (1998) The regulation of E2F by pRB-family members. *Genes and Development* **12**, 2245–2262.

Fearon E.R. (1998) Human cancer syndromes: clues to the origin and nature of cancer. *Science* **278**, 1042–1050.

Hanahan, D. & Weinberg, R.A. (2000) The hallmarks of cancer. *Cell* **100**, 57–70.

Hunter, T. (1997) Oncoprotein networks. *Cell* **88**, 333–346.

Lane, D. (1998) Awakening angels. *Nature* **394**, 616–617.

Lengauer, C., Kinzler, K.W. & Vogelstein, B. (1998) Genetic instabilities in human cancers. *Nature* **396**, 643–647.

Levine, A.J. (1997) p53, the cellular gatekeeper for growth and division. *Cell* **88**, 32–33.

Paulovich, A.G., Toczyski, D.P. & Hartwell, L.H. (1997) When checkpoints fail. *Cell* **88**, 315–321.

Apoptosis

Adams, J.M. & Cory, S. (1998) The Bcl-2 protein family: arbiters of cell survival. *Science* **281**, 1321–1326.

Ashkenazi, A. & Dixit, V.M. (1998) Death receptors: signalling and modulation. *Science* **281**, 1305–1309.

Green, D.R. & Reed, J.C. (1998) Mitochondria and apoptosis. *Science* **281**, 1309–1312.

Jacobson, M.D., Weil, M. & Raff, M.C. (1997) Programmed cell death review in animal development. *Cell* **88**, 347–354.

Reed, J.C. (1997) Double identity for proteins of the Bcl-2 family. *Nature* **387**, 773–777.

Rich, T., Watson, C.J. & Wyllie, A. (1999) Apoptosis: the germs of death. *Nature Cell Biology* **1**, E69–E71.

Index

Page numbers in *italic* refer to figures and those in **bold** refer to tables.